ordering lives

AN INTRODUCTION TO THE SOCIAL SCIENCES: UNDERSTANDING SOCIAL CHANGE

This book is part of a series produced in association with The Open University. The complete list of books in the series is as follows:

Questioning Identity: Gender, Class, Nation
edited by Kath Woodward

The Natural and the Social: Uncertainty, Risk, Change
edited by Steve Hinchliffe and Kath Woodward

Ordering Lives: Family, Work and Welfare
edited by Gordon Hughes and Ross Fergusson

A Globalizing World? Culture, Economics, Politics
edited by David Held

Knowledge and the Social Sciences: Theory, Method, Practice
edited by David Goldblatt

The books form part of the Open University course DD100 *An Introduction to the Social Sciences: Understanding Social Change*. Details of this and other Open University courses can be obtained from the Course Reservations Centre, PO Box 724, The Open University, Milton Keynes MK7 6ZS, United Kingdom: tel. +44 (0)1908 653231, e-mail ces-gen@open.ac.uk

Alternatively, you may visit the Open University website at http://www.open.ac.uk where you can learn more about the wide range of courses and packs offered at all levels by The Open University.

For availability of other course components, contact Open University Worldwide Ltd, The Berrill Building, Walton Hall, Milton Keynes MK7 6AA, United Kingdom: tel. +44 (0)1908 858785; fax +44 (0)1908 858787; e-mail ouwenq@open.ac.uk; website http://www.ouw.co.uk

ordering lives:
family, work and welfare

edited by gordon hughes and ross fergusson

London and New York

in association with

The Open
University

First published 2000 by Routledge; written and produced by The Open University
11 New Fetter Lane, London EC4P 4EE

Simultaneously published in the USA and Canada by Routledge
29 West 35th Street, New York, NY 10001

Routledge is an imprint of the Taylor & Francis Group

Edited, designed and typeset by The Open University.

Printed by The Bath Press, Bath.

British Library Cataloguing in Publication Data
A catalogue record for this book is available from The British Library

Library of Congress Cataloging in Publication Data
A catalogue record for this book has been requested

ISBN 0-415-22291-5 (hbk)

ISBN 0-415-22292-3 (pbk)

1.1

Contents

The Open University course team

John Allen, *Senior Lecturer in Geography*

Penny Bennett, *Editor*

Pam Berry, *Compositor*

Simon Bromley, *Senior Lecturer in Government*

David Calderwood, *Project Controller*

Elizabeth Chaplin, *Tutor Panel*

Giles Clark, *Co-publishing Advisor*

Stephen Clift, *Editor*

Allan Cochrane, *Professor of Public Policy*

Lene Connolly, *Print Buying Controller*

Graham Dawson, *Lecturer in Economics*

Lesley Duguid, *Senior Course Co-ordination Secretary*

Ross Fergusson, *Staff Tutor in Social Policy*

Fran Ford, *Senior Course Co-ordination Secretary*

David Goldblatt, *Co-Course Team Chair, Lecturer in Government*

Jenny Gove, *Lecturer in Psychology*

Judith Greene, *Professor of Psychology*

Montserrat Guibernau, *Lecturer in Government*

Peter Hamilton, *Lecturer in Sociology*

Celia Hart, *Picture Researcher*

David Held, *Professor of Politics and Sociology*

Susan Himmelweit, *Senior Lecturer in Economics*

Steve Hinchliffe, *Lecturer in Geography*

Gordon Hughes, *Lecturer in Social Policy*

Christina Janoszka, *Course Manager*

Pat Jess, *Staff Tutor in Geography*

Bob Kelly, *Staff Tutor in Government*

Margaret Kiloh, *Staff Tutor in Applied Social Sciences*

Sylvia Lay-Flurrie, *Secretary*

Siân Lewis, *Graphic Designer*

Tony McGrew, *Professor of International Relations, University of Southampton*

Hugh Mackay, *Staff Tutor in Sociology*

Maureen Mackintosh, *Professor of Economics*

Eugene McLaughlin, *Senior Lecturer in Applied Social Science*

Andrew Metcalf, *Senior Producer, BBC*

Gerry Mooney, *Staff Tutor in Applied Social Sciences*

Ray Munns, *Graphic Artist*

Kathy Pain, *Staff Tutor in Geography*

Clive Pearson, *Tutor Panel*

Lynne Poole, *Tutor Panel*

Norma Sherratt, *Staff Tutor in Sociology*

Roberto Simonetti, *Lecturer in Economics*

Dick Skellington, *Project Officer*

Brenda Smith, *Staff Tutor in Psychology*

Mark Smith, *Lecturer in Social Sciences*

Grahame Thompson, *Professor of Political Economy*

Ken Thompson, *Professor of Sociology*

Stuart Watt, *Lecturer in Psychology/KMI*

Andy Whitehead, *Graphic Artist*

Kath Woodward, *Co-Course Team Chair, Staff Tutor in Sociology*

Chris Wooldridge, *Editor*

External Assessor

Nigel Thrift, *Professor of Geography, University of Bristol*

Series preface

Ordering Lives: Family, Work and Welfare is the third in a series of five books, entitled *An Introduction to the Social Sciences: Understanding Social Change*. If the social sciences are to retain and extend their relevance in the twenty-first century there can be little doubt that they will have to help us understand social change. In the 1990s an introductory course to the social sciences would have looked completely different.

From a global perspective it appears that the pace of change is quickening, social and political ideas and institutions are under threat. The international landscape has changed; an intensification of technological change across computing, telecommunications, genetics and biotechnology present new political, cultural and moral dilemmas and opportunities. Real intimations of a global environmental crisis in the making have emerged. We are, it appears, living in an uncertain world. We are in new territory.

The same is also true of more local concerns. At the beginning of the twenty-first century both societies and the social sciences are in a state of flux. *Understanding Social Change* has been written at a moment that reflects, albeit in a partial way, subterranean shifts in the social and cultural character of the UK. Established social divisions and social identities of class, gender, ethnicity and nation are fragmenting and re-forming. Core institutions such as the family, work and welfare have become more diverse and complex. It is also a moment when significant processes of change have been set in train – such as constitutional reform and European economic and monetary union – whose longer-term trajectory remains uncertain. The flux in the social sciences has been tumultuous. Social change, uncertainty and diversity have rendered many of the most well-established frameworks in the social sciences of limited use and value. Social change on this scale demands fresh perspectives and new systems of explanation.

In this context *Understanding Social Change* is part of a bold and innovative educational project, for it attempts to capture and explore these processes of momentous social change and in doing so reasserts the utility and necessity of the social sciences. Each of the five books which make up the series attempts precisely this, and they all do so from a fundamentally interdisciplinary perspective. Social change is no respecter of the boundaries of disciplines and the tidy boxes that social scientists have often tried to squeeze it into. Above all, *Understanding Social Change* seeks to maintain and extend the Open University's democratic educational mission: to reach and enthuse an enormously diverse student population; to insist that critical, informed, reflexive engagement with the direction of social change is not a matter for elites and professional social scientists alone.

As you may have guessed, this series of books forms a core component of the Open University, Faculty of Social Sciences, level 1 course, DD100 *An Introduction to the Social Sciences: Understanding Social Change*. Each book in the series can be read independently of the other books and independently from the rest of the materials that make up the Open University course. However, if you wish to use the series as a whole, there are a number of references to chapters in other books in the series, and these are easily identifiable because they are printed in bold type.

Making the course and these books has been a long and complex process, and thanks are due to an enormous number of people.

First and foremost, the entire project has been managed and kept on the rails, when it was in mortal danger of flying off them, by our excellent Course Manager, Christina Janoszka. In the DD100 office, Fran Ford, Lesley Duguid and Sylvia Lay-Flurrie have been the calm eye at the centre of a turbulent storm, our thanks to all of them.

Stephen Clift, Chris Wooldridge and Penny Bennett have been meticulous, hawk-eyed editors. Siân Lewis has provided superb design work, and Ray Munns and Andy Whitehead have provided skilled cartographic and artistic work. David Calderwood in project control has arranged and guided the schedule with calm efficiency and Celia Hart has provided great support with illustrations and photographs. Nigel Thrift, our external assessor, and Clive Pearson, Elizabeth Chaplin and Lynne Poole, our tutor panel, have provided consistent and focused criticism, support and advice. Peggotty Graham has been an invaluable friend of *Understanding Social Change* and David Held has provided balance, perspective and insight as only he can.

It only remains for us to say that we hope you find *Understanding Social Change* an engaging and illuminating introduction to the social sciences, and in turn you find the social sciences essential for understanding life in the twenty-first century.

David Goldblatt
Kath Woodward
Co-Chairs, The Open University Course Team

Introduction

Ross Fergusson and Gordon Hughes

Imagine a world without routine or without order, where nothing stays the same, and where there is little that does not change from moment to moment, or from year to year. But this is not a description that accurately depicts many people's lives or much of the social world in which we all live. Lives and societies are predictable. In many ways, perhaps remarkably so. Most people live in families of one kind or another; most adults go out to work on a regular basis until they retire. Patterns repeat themselves and stability rather than chaos is the mark of everyday life. But why should this be? How are we to make sense of the ways in which our lives are routinely ordered, while *also* recognizing that change is a prominent feature of contemporary society? These are the questions that this book sets out to answer.

Social scientists argue that societies are in large measure held together through their key social institutions. They order lives both by shaping our daily activities and by defining many of the possible ways in which individuals and groups relate to one another. In both senses, ordering lives entails the exercise of power. Through institutions, the preferences, aims and interests of some groups and individuals are served while those of others are given less priority or neglected altogether. However, the patterns of power within social institutions are not fixed, nor are the kinds of ordering they foster. Institutions change, and sometimes these changes lead to transformations in the way individuals' lives and whole societies are ordered.

To help us in this exploration of the changing patterns of everyday life, we focus on three central institutions of UK society: *the family, work and employment,* and *the welfare state,* in terms of ordering lives and ordering the relations between people. The book looks at how power works in and through these institutions, and at some recent transformations in ordering them. As the chapter titles indicate, transformations in all three institutions represent significant social change. The ways in which UK society is held together and lives are ordered is clearly changing rapidly. But how much and with what effects can only be fully understood by looking at how power used to work and how it works now: in our case in families, in work and employment, and through welfare.

Ordering lives

The idea of 'ordering lives' expresses the importance of processes in understanding the diverse ways in which our lives are organized. It also helps us understand how institutions hold together yet also change *over time.*

Throughout the book you will find the term 'ordering' used much more than the more familiar term 'order' because we want to emphasize that ordering is a *continuous* process. If we talk about 'the order of your life', or even something as grand as 'the social order' it implies that there is one, relatively fixed order. And although some aspects of order seem to be remarkably long-lasting, even the most fundamental and taken-for-granted kinds of order do change.

Changing social institutions

Social institutions are, in part, about order. They exist to create particular kinds of order. They come into being because of their potential for ordering crucial aspects of our lives. We all live in and have our lives ordered by a vast array of institutions that take many different forms. In some cases these are physical institutions with discernible buildings and relatively strict regimes for ordering lives: schools, hospitals, government offices, places of work and such like. But the concept of an institution has a much broader meaning. It not only embraces tangible 'everyday' institutions like the family, but also more distant offices of the state like the Cabinet, the market in jobs, and even spheres of activity like the United Nations, the BBC or CNN. Although very different from each other in many ways, all these institutions involve sets of relations that are circumscribed or bounded. What all institutions have in common is that:

- they are made up of a predictable set of social practices
- they are usually governed by clear procedures, norms or regulations
- some important aspects of social, economic or cultural life are conducted and ordered through them.

The three institutions (family, work and welfare) on which this book primarily focuses all have these qualities in abundance: they are predictable, they are governed, and they order lives. But in all these respects they are also changing. Indeed, all three have undergone transformations in the activities and practices they govern and order. In each of them, the two decades immediately after the Second World War are often characterized as a 'golden age' of stability, security, certainty, well-being and order. The key transformations are those that occurred during the last quarter of the twentieth century and brought this supposed 'golden age' to an end.

Uncertainty and diversity

The transformations in the family, in work and in welfare have all been profound, unimaginable even, when viewed from the securities of life in the immediate post-Second World War period in the UK. To some, the transformations have overturned the stable, ordered life of that era, and put in its place uncertainty, insecurity and threats to the order that holds UK society together. To others, these transformations have lifted the veil on a 'golden age' that was 'golden' only for some. They have broken the oppressive mould of predictable, routinized lives and standard ways of living, and in doing so have fostered diversity and opened opportunities.

In the family, the traditional nuclear family form, with its circumscribed gender roles and its taken-for-granted set of domestic living arrangements has been challenged, even abandoned by some, in favour of other ways of living together and bringing up children. Chapter 2 traces these changes. It looks at the competing analyses, views and values of those who see this transformation as a hazardous leap into an uncertain future, and those who view it as the end of old injustices and inequalities, and the beginning of bright new opportunities and a rich diversity of living arrangements.

In work, the old certainties of secure employment, a 'job for life', and a reliable wage have been eroded for some, to be replaced by flexible forms of working. Chapter 3 begins by discussing the emergence of secure jobs in an age of full male employment in the decades directly after the Second World War, and the subsequent transformation into diverse new arrangements of temporary and part-time work, with new winners and new losers.

In welfare, the security and well-being offered by an extensive welfare state have been labelled as unaffordable and a disincentive to enterprise, and are being reworked into arrangements that rely more on personal responsibility and private provision. Chapter 4 looks at the history of this transformation, and at the powerfully contested views of whether it represents the end of security for all and an end to equal access to essential forms of welfare, or whether it marks a vigorous new era of diverse arrangements for self-reliance and initiative, unencumbered by the burdens of a debilitating welfare state.

Power and theory

How are we to make sense of these transformations? Are they sources of uncertainty and insecurity that threaten the kinds of order that hold UK society together? Or do they mark the end of old restrictions and the beginning of flourishing new ways of living and working that are creating new kinds of order? To answer these questions we need a clearer understanding of how institutions order people's lives. How much are the ordered routines of our lives chosen by us? How much are they forced on us by a particular set of interests? Do different kinds of order favour different individuals and groups? The concept of power is essential to answering these questions, and we need clearly reasoned explanations – theories – of the ways in which power works to make use of this concept.

Behind order, then, there is power – power to decide why this kind of order is better than that kind. And, as we will be showing, power is not fixed forever. Power changes. Sometimes one group of people or one institution is more powerful, sometimes another. As every news bulletin implicitly tells us, the publicized events of every day can be understood as part of a constant struggle to redefine the boundaries of power. At its most large-scale and extreme, wars between nation-states are struggles for power. But much more commonplace and less dramatic are the power struggles between politicians, bureaucracies, campaigners and so on. In turn, firms struggle for power in the market. Even more familiar and mundane are the small-scale negotiations, stand-offs and conflicts of people's everyday lives at work, at home, in playgrounds and in pubs. Some are so low-key, small-scale and chaotic that they are hardly identifiable as struggles, or as being about power. But behind many of these encounters is one person's wish to get another to act or think differently. So power, like 'order', is being continuously used, contested and, some would say, redefined.

How does the perpetual and shifting exercise of power order people's lives? Can people truly be forced to act in particular ways? Do they choose to accept that these are the only ways to act? Or are there other ways of understanding power? How did the transformations in the family, in work and in the welfare state come about? How did the lines of power move to allow such profound changes? To begin answering questions like these, some theories are needed to analyse how power works, and develop reasoned explanations of it. Beginning in Chapter 1, two very different theories, put forward by the German sociologist Max Weber, and by the French philosopher, Michel Foucault, are used to help think through the ordering of lives in and through institutions. Is power, as Weber suggests, something that certain people 'hold' and 'exercise' over others, or, following Foucault, is it more valuable to see power as something that 'works through' institutions and agents?

Even the simplest of social changes must, in the end, be imposed or emerge through extensive processes of negotiation as newly accepted ways of doing things. We explain in Chapter 1 how, in a society as complex as ours, a change as relatively simple as altering the kinds of food available to us entails the exercise of power. The campaigns, the protests, the parliamentary battles and the violent conflicts over food risks that have accompanied the endeavours of giant food-producing corporations to introduce genetically modified foods provide us with one example that has demonstrated this graphically. The power of multi-national companies, the power of consumers, the power of activists and the power of mass media to transmit and shape ideas all vie to secure wide acceptance or outright rejection of change. Are coercion, persuasion or domination effective ways of securing change? What understandings of how institutional power works in firms and in government can be used to make sense of how one powerful party (GM food suppliers) strives to prevail over another (sceptical consumers)? The insights of Weber's and Foucault's theories offer sharply contrasting ways of understanding these processes. Together with Marxist theories of power, they provide theoretical frameworks to analyse how power operates in the more complex processes of ordering people's lives in families, through work, and through welfare as discussed in Chapters 2–4.

Political ideologies

Theories of power are all concerned with how particular ideas, beliefs and values about what is necessary and what is desirable come to hold sway. Some of these ideas are so recurrent and coherent as to form identifiable clusters of beliefs and values that make sense of the world, and profoundly shape our actions as a result.

These political ideologies are centrally concerned with power, and how it is exercised to promote the ends they value. They have also driven some of the key changes in thinking that have transformed the ways in which our lives are ordered in families, how work and employment are ordered and order us, and how state welfare shapes our lives.

When we look at the family in Chapter 2, we use the political ideologies of conservatism and feminism to compare ways of explaining some of the foundations of the 'traditional' family and some of the changes to family life that have occurred. In Chapter 3, our explanation of how work and employment has been re-ordered focuses first on the influence of social democracy in creating full (male) employment and 'jobs for life', and then on liberalism's influence on the creation of flexible employment markets. In the final chapter, we draw together several ideologies to understand how the welfare state came to be set up, and how it is changing.

As you read the four chapters of this book you will encounter in greater depth the broad questions and themes outlined in this Introduction. By the end of the book, you should be ready to revisit these questions and themes, and to explore, in a more informed and critical way, possible future transformations in the institutional ordering of our lives that may occur in the twenty-first century. We hope you enjoy the intellectual journey.

Power: its institutional guises (and disguises)

John Allen

chapter 1

Contents

1 INTRODUCTION

The main purpose of this chapter is to explore what power *is* and how it *works* so that, throughout the book, we are in a position to examine what holds together the changing institutions of the family, work, and social welfare to which many of us are attached. In particular, a major concern is to show how power in its various guises serves to integrate and stabilize these key areas of social life, although often in less certain ways than many would have us believe. That such institutions are always in the process of being ordered in one way or another, rather than simply fixed for all time, is something that you should bear in mind as you consider the nature of *institutional* power.

By institutional power what we have in mind are the relatively routine arrangements of power involved in, say, the bringing up of children within a family, the management of an office within a large company, or indeed the exercise of judgement by government officials over what is fit and proper to eat, or over who is deserving of welfare and who is not. As can be gleaned from these few instances, the issue of power is not one restricted to the likes of benevolent dictators, American Presidents or military commanders; it is something that affects us all simply because we are part and parcel of this ordered set of relationships that we call 'society'. To that end, in all kinds of ways we may find ourselves immersed in the many arrangements of institutional power, whether we recognize them or not.

Perhaps it is useful at the outset to state that institutional power and the quest for order is, for most of us, experienced as something quite *ordinary*: it is often, for example, the kind of thing that you only really know about when you are on the receiving end of it. When as a child you may have complied with or rebelled against the discipline laid down by a parental figure or a teacher, or later in life found yourself following, perhaps against your will, the instructions given by a bureaucratic manager, a doctor, or some other notable authority, the encounter, broadly speaking, may be described as a brush with power. In particular, you may have found yourself doing something which you did not really set out or want to do. The experience of imposition, of not being entirely in control of your own actions, is a familiar, indeed ordinary, consequence of finding yourself on the receiving end of an act of power. In such circumstances, it may seem relatively clear as to who has the power and who does not, even if the chain of command starts elsewhere in some far-off government ministry or in the distant headquarters of a foreign multinational.

It is no less ordinary, however, to experience institutional power in far more anonymous ways, where it is altogether unclear that anyone is directing or controlling things.

ACTIVITY 1.1

This sense of power is not so unfamiliar as it might at first appear. It is a rather different brush with power from that of the authority figure: less direct, more impersonal and rather difficult to discern who is behind it. Take a minute or two to consider or recall any circumstances where you may have felt on the receiving end of some kind of pressure to act or to order your life in a certain way, at work perhaps, or in a public place, yet found it hard to say why exactly.

FIGURE 1.1   Closed-circuit television cameras: an anonymous brush with power?

Perhaps the obvious examples here are those of closed-circuit television cameras placed discreetly in public locations such as shopping malls or the computerized personal data banks held by governments, both of which, in different ways, give the impression that your behaviour is subject to some form of scrutiny and possibly reproach. Surveillance, none the less, is a known form of control, even if you cannot pinpoint precisely the location of that power. The same, however, cannot be said about the power associated with the likes of advertising and fashion. Here we may experience power through a particular imagery which is suggestive of what it is to be this or that kind of person. No prohibitions are ever laid down for us as to how we should be, yet some people – knowingly or without much thought being given – seem to mould themselves along the lines of current fashion, for example in relation to a particular 'healthy' lifestyle or a certain 'desirable' type of body.

This type of issue should strike a familiar chord in relation to the topic of social identity (see **Woodward, 2000**). No one, after all, tells you to be who you are, yet there is a sense in which people may condition themselves, perhaps discipline themselves even, to appear 'normal' – whatever that may mean. With this kind of example, however, we are in danger of running ahead of ourselves before we have had an opportunity to tease out the obvious from the less obvious experiences of power and order in society.

What we can draw from the above, however, is an elementary definition of power:

> A relationship of power is evident when *someone acts in a way which they would otherwise not have done* – regardless of whether or not they chose to.

Let's be clear about what is meant by elementary here. This is an elementary definition of power because it represents the *starting point*, not the end point, of the analysis. Much of this chapter will be spent exploring the ways in which different attempts to order our lives are put together – cultural, political and economic – with an eye to constructing a more elaborate view of power. In the next two sections I develop a specific example (which connects with **Hinchliffe and Woodward, 2000**) – the debate around 'safe' food at a time when genetic engineering promises the earth – to draw out the various modes of institutional power in play: the role of persuasion, the reliance on those in authority, the practice of domination, and so forth. Following that, in Sections 4–6, I spell out in detail two contrasting theories of power to show how it is possible to think of power as, either something which is held *directly* over others, or as something which passes *indirectly* through the hands of the powerful no less than through the hands of the powerless. Where power in the latter view seeks to establish order by working *on* peoples' actions and beliefs, in the former it is exercised *over* them. Finally, in Section 7, I look briefly at the role of theory in constructing explanations of events.

For the moment, however, we remain with the highly charged issue of genetically-modified (GM) foods and the vast implications that it appears to carry for what we eat and how we live. As you work your way through the issue, keep in mind the elementary definition of power as we shall return to it in Section 3 in relation to the foods that we consume.

2 FOOD FOR THOUGHT

We are what we eat, or so we once thought. The concern voiced over the application of genetic engineering to foods has taken an odd twist in recent years. In the late 1990s, self-styled 'eco-warriors' appeared on the UK agricultural scene destroying dozens of genetically-modified crop trials at test sites up and down the country. Dressed in protective clothing, reputedly often masked, groups of activists under the cover of darkness apparently stormed the target fields and bagged what they saw as the offending crops. Some groups gave themselves suitable sounding names, such as the 'Wardens of Wiltshire', the 'Lincolnshire Loppers', the 'Kenilworth Croppers', or more imaginatively, the 'Super Heroes Against Genetix'. Among the main crops targeted for destruction were fields of genetically-modified soya, sugar beet, wheat and oil seed rape, with the faceless biotechnology multinationals held up as the villains of the piece for their rash and irresponsible behaviour in pushing forward the boundaries of genetic manipulation at such a reckless pace.

FIGURE 1.2 Direct protest against genetically-modified crops: the rolling action of GenetiX Snowball campaigners

Direct protest, however, against the appearance of what many activists regarded as 'Frankenstein foods' was about to move out of the fields into the supermarkets. Demonstrations outside of supermarkets which stocked genetically-altered products – many processed foods in fact – had people turning up to protest dressed as vegetables crossed with chickens to forcefully make the point about genetic distortion. On other occasions, protesters stacked their supermarket trolleys to the rim with assorted foods and then demanded of the checkout staff that they be told how much genetically-modified material the products contained.

Nor did the matter stop there. Organic farmers, perhaps rather predictably, were up in arms for fear of cross-pollination between the 'mutant' crops and their own 'pure' product. Iceland, the domestic frozen food retailer with a sizeable chunk of the UK food market, came out firmly against genetically-modified foods and banned all GM material from its own-brand products. In the wake of public outcry over the spectre of tampered foodstuffs, other supermarkets were equally quick to review their product lines. Even the House of Commons all-party catering committee took the decisive step to bar all genetically-altered foods from the Commons restaurants, despite the fact that the Labour government of the day was sharply divided over the issue. And to top it all, Prince Charles threw his hat into the ring by expressing grave concern over genetic modifications and declaring that some realms are best left to God, and to God alone.

What are we to make of all this in terms of the power relationships involved? First things first: before we move on to consider the different groups, organizations, and authorities involved, as well as their varied interests, what was all the fuss about?

There have been a number of concerns expressed about the introduction of genetically-modified materials into the food chain. At one level, concern reflects the well-founded anxiety surrounding the foods that we eat. The danger to health that such foods may represent, the uncertainties attached to any form of genetic experimentation, are sufficient in themselves to warrant concern over the possible risks involved. Opinion is divided, of course, as to the extent of the risks involved, but such worries are compounded by the fact that it seems far from easy to detect the presence of GM material in many finished products. This may change over time, but so long as, for example, the majority of processed foods available in the UK – those such as biscuits, bread, chocolate and cakes – contain soya, and much of that soya comes from the USA where the identification of the GM component is not considered an issue, then such concerns are likely to remain. If consumers have no choice over what they eat, nor full knowledge of the potential risks, the anxiety – following panics like that of the BSE scare – is real enough one would have thought (**Hinchliffe, 2000**).

It is at the broader environmental level, however, that the risks attached to genetically-engineered crops have stimulated the most animated responses. The fear, quite simply, is that the widespread use of crops which carry artificially-introduced genes could seriously upset the ecological balance. Once new genes escape into the wild they cannot be rounded up and led back. They represent an unknown quantity, with, so the environmental opposition claims, potentially devastating consequences for all forms of wildlife. In the long term, for example, crops endowed with insect-killing genes may harm beneficial insects or upset the habitat of birds and insects alike. Such crops, moreover, because of their engineered resistance to certain herbicides may give rise to a new breed of 'superweed' which could pose a threat to many conventional crops.

The 'grey areas' of knowledge are extensive, with hard-nosed proponents of genetic engineering dismissing arguments based on the superficial fear of interfering with 'nature' as melodramatic. Genetic engineering, they argue, is nothing more than an extension of common breeding techniques currently applied to plants and animals, and 'nature' quite frankly is little more than the accumulated technological adaptations of previous generations. The ability to feed future generations and to make inroads into malnutrition and famine in the less developed world rests with exploiting the benefits of scientific progress. And so the argument rolls on.

The precise detail of the arguments mobilized by the different camps is not our major concern here. At this stage, you need only be aware of the main lines of disagreement. These are summarized in Box 1.1, 'Whose choice is it anyway?'

BOX 1.1   Whose choice is it anyway?

Genetic modification is the process whereby genes controlling specific characteristics in plants or animals are isolated and transferred to other organisms. For example, the American biotechnology company Monsanto has developed soya beans modified with genetic material from soil bacteria, petunias and a virus to make them herbicide tolerant. The imported genetic material is fired into DNA.

Genetically modified soya, tomatoes, yeast, oilseed rape, and maize have been approved for use in the UK.

Most processed foods – everything from bread and biscuits to baby food and beer – contain soya in some form.

Experts have argued for and against the spread of genetically modified foods.

Those for say:

- genetic engineering can improve the quality of food, by making it last longer, improving its flavour
- it could make food healthier by altering its nutritional content, for example, by reducing levels of fat
- it could help feed the world by improving yields and reducing costs
- fewer pesticides and herbicides will be needed, helping soil conservation; chemicals can be better targeted; less energy will be used in spraying.

Those against say:

- scientists cannot anticipate potential risks, as BSE shows
- it will encourage monoculture of crops which are potentially susceptible to disease
- it concentrates too much power in the hands of agrochemical giants
- it could lead to a loss of biodiversity
- consumers are being given no choice, without knowing the consequences
- the technology poses fundamental ethical questions about life forms
- there have been disturbing new toxins and allergies related to GM foods.

(From *The Guardian*, 4 June 1998, p.15.)

ACTIVITY 1.2

Cast your eye over what those 'for' and those 'against' have to say on the issue of GM foods (Box 1.1). As you read through the general description in Box 1.1, take note of the last sentence. It points out that, *experts* 'have argued for and against the spread of genetically modified foods'.

Now we know something about those who have voiced their concerns and misgivings about this latest modification of the foods that we eat – the so-called 'eco-warriors', the environmental activists, consumer protesters, organic farmers and the like – as well as something about those promoting it such as the biotechnology companies, who do you think are the *experts*?

2.1 Experts in the field

The short answer to this question is that the experts are drawn from the scientific community which works on genetic issues, but this observation is perhaps less helpful than it seems. There are scientists employed by all the big multinationals involved in biotechnology, such as Monsanto, the US leader already mentioned, Novartis, Europe's largest pharmaceuticals company, and Astra Zeneca, a major Swedish/UK drugs and agrochemicals operation. There are scientists who work in government-sponsored research establishments who conduct trials with the major companies or independently of them. Members of the latter will belong to various committees of experts which advise government on genetic releases or on matters of environmental regulation more generally. The Soil Association has its own experts to call on, as do Greenpeace, Friends of the Earth, and any number of interested organizations from the National Farmers' Union and the RSPB to Christian Aid and Oxfam.

If there is a scientific community out there which concerns itself with questions of genetics, then it is an especially diverse one. So, let's start with the company scientists and the manner in which they are attempting to influence the outcome of the food debate.

The positive assessment of the risks attached to genetic engineering given by the scientists from companies such as Monsanto rests upon their extensive field trials. Figure 1.3 provides a recent indication of some of the sites in England, Scotland and Wales where Monsanto and other biotechnology firms and agencies have tested genetically-modified crops. In all, they are tiny in number when set alongside the global spread of Monsanto's field trials, said to involve more than 40 countries and around 60 different crops. With an official approval process in the UK that could take up to three years before genetically-altered foods reach the market, the weight of their case is obviously related to the success of the field trails. It is hardly surprising then, to witness the company's intense irritation with the destructive acts committed by the likes of the 'Kenilworth Croppers' who, plainly, are pulling up the very evidence that Monsanto is reliant upon to gain widespread approval.

1 Aberdeenshire
MacRobert Farm, Aberdeen 1
West Balhalgardy, Inverurie 1
Scottish Agricultural College,
  Tillycorthie Farm, Udny 8
Scottish Agricultural College,
  Craibstone Estate, Bucksburn 2

2 Berkshire
Jealott's Hill Research Station,
  Bracknell 2
plus farm scale trials in Reading

3 Cambridgeshire
National Inst. of Agricultural
  Botany, Cambridge 17
Bannold Road, Waterbeach 2
Plant Breeding Int., Trumpington,
  Cambridge 1
Sacrewell Lodge Farm,
  Peterborough 2
Svalof Wiebull Ltd, Huntington 2
Further Fen Farm, Southery 1
Worlick Farm, Forty Foot 1
Morley Farm, Warboys 1
New Farm, Cambridge 1

4 Derbyshire
Wilson Hall Farm, Melbourne 1

5 Essex
Rainham Farm, Rainham 1

6 Gloucestershire
Laverton Meadows,
  Nr Broadway 1

7 Hampshire
ADAS Bridgets, Winchester 6
National Inst. of Agricultural
  Botany, Winchester 2

8 Hertfordshire
Summerhouse Farm, Royston 4
Vine Farm, Nr Royston 2
IACR Rothampsted,
  Harpenden 1
Chishil Orchard Farm,
  Royston 2
Tilekiln Farm,
  Bishops Stortford 1
Wood Farm,
  Hemel Hempstead 1
plus farm scale trials in Harpenden
  and Hemel Hempstead

9 Lincolnshire
Dalby Farms Ltd, Brigsley 7
GH Hoyles Ltd, Spalding 1
Barr Farm, Horncastle 2
Advanta Seeds UK Ltd,
  Lincoln 2
Sharpes Int. Seeds Ltd, Lincoln 1
Elsoms Seeds, Spalding 1
Monmouth House, Spalding 2
Home Farm, Glentham 1
Plant Breeding Station,
  Boothby Graffoe 1
The Old Rectory,
  Market Rasen 1
plus farm scale trials in Boothby
  Graffoe, Market Rasen and
  Glentham 2

10 Lothian
Scottish Agricultural College, Boghall Farm,
  Edinburgh 5

11 Norfolk
Morley Research Centre, Wymondham 7
Wannage Farm, Downham Market 2
John Innes Centre, Colney 1
Manor Farm, Bradenham 1
East Winch Hall, King's Lynn 1
The Grange, King's Lynn 1
plus farm scale trials in Norwich

12 North Yorkshire
Headley Hall Farm, Tadcaster 3
ADAS High Mowthorpe, Malton 2
University of Leeds, Tadcaster 1

13 Nottinghamshire
ADAS Gleadthorpe Research Centre,
  Mansfield 2
Top Brackendale, Bingam 1
Home Farm, Bingam 1
plus farm scale trials in Bingham

14 Oxfordshire
Hills Nurseries, Abingdon 1
Model Farm, Watlington 1
University Farm, Oxford 1
plus farm scale trials in Watlington

15 Perthshire
East Loan Field, Invergowrie 1

16 Shropshire
Eyton House Farm, Telford 3

17 Somerset
IACR Long Ashton Research Station,
  Somerset 1

18 Suffolk
Brooms Barn Experimental Station,
  Bury St. Edmunds 1
Hall Farm, Bury St. Edmunds 2
IACR Brooms Barn, Bury St. Edmunds 3
Street Farm, Denham 1

19 Worcestershire
Thorn Farm, Inkberrow 3
East Lodge Farm, Broadway 3

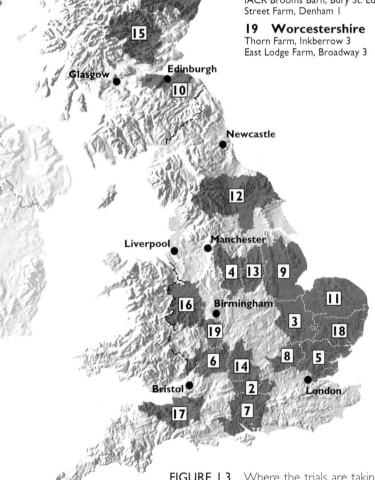

FIGURE 1.3 Where the trials are taking place as at 17 August 1999. Figure after each location indicates number of licences

Source: *The Guardian*, 17 August 1999

Government scientists in the UK seem to vary in the extent of their enthusiasm for the GM cause depending on the ministry in question, whilst the deepest reservations about the new genetic technologies have come from a wide range of authoritative sources. One of the key concerns expressed is that the whole introduction of genetically-modified seeds and foods has happened far too quickly. Corporate interests, a number of environmental scientists have argued, are dictating the pace of development, not people's futures. It is one thing to point to extensive geographical tests, it is quite another to say that sufficient time has been given to the screening of all possible risks to human kind and the ecological system at large. The issue is just too big to gamble with – especially when the risks taken may affect us all.

The language on all sides is uncompromising, the 'grey areas' of knowledge as indicated are extensive, and feelings on the issue often run high. So where do the lines of power run on the issue? To what extent are consumers powerless to resist the (often unidentifiable) march of genetically-modified foods on to the supermarket's shelves? How effective is direct action against the might and resources of the big biotechnology companies? How powerful is the scientific opposition? Who has the authority on such matters? Will we be forced to eat genetically-modified foods whether we like it or not?

It is to these and other related questions of power that we now turn.

3   POWER AT WORK

Coercion
To compel by force or its threatened use.

In response to the last question we can answer quite emphatically: no. There may be a matter of restricted choice involved, but force is not the issue here. Leaving to one side the obvious point that nobody will be physically forced to eat GM foods, the threat of force, or rather the use of **coercion**, is an equally unlikely occurrence. Coercion is hardly suited to the matter of winning people over to the benefits of certain foods, nor is the threat of force particularly effective over time. The meaning of force in the question above is decidedly loose and, from the point of view of those who oppose the genetic alteration of foods, cajolement is probably closer to their understanding of the issues. In this case, they would very likely accuse the big producers of **manipulation**; that is, of hiding from the public the dangers that they believe to be present in the altered foods. On this view, consumers are being duped, or worse, led without their knowledge to act as guinea pigs in some sort of grand food experiment.

Manipulation
To conceal the real intent behind an action in order to gain an advantage.

Whether the suggestion is true or false is not the issue here; rather, the point concerns the kind of power that is actually being wielded. In order to understand how power *works*, we first need to know something about the different modes through which it is exercised. By *modes*, I simply mean the various forms that power takes in society. Coercion, for instance, is a distinct mode of power, quite different in fact from manipulation and, as we shall see, quite different from either authority or domination. If we mix up the various modes, we lose any sense of the different circumstances to which they correspond.

So, when a large multinational institution such as Monsanto draws upon its resources to achieve certain ends, in what sense could it be said to have acted in powerful ways? If coercion or manipulation is not part of its armoury, for instance, has the company sought to dominate the marketplace and constrain the voice of consumer opposition? Or, if this is far too strong a language to describe its mode of operation, has the company sought to stamp its expertise and authority on all matters concerning genetically-modified foods? Or, failing that, has Monsanto merely acted to persuade the consuming public at large of the merits of such foods and their general, all-round safety?

We already know a little about some of these matters, so let's take a closer look at the arrangements of power put together by Monsanto and other biotechnology multinationals. In order to focus the enquiry, it is worthwhile to consider it in terms of the elementary definition of power stated earlier, namely:

> Have the biotechnology companies exercised their power in such a way that *the public have consumed foods which they would otherwise not have done?*

3.1   If coercion is not the answer ...

Let's start by looking at the act of **persuasion**.

Back in the late 1990s, Monsanto launched a one million pound advertising campaign to win the 'hearts and minds' of the British consuming public. Over a three month period, the company placed a series of full page advertisements in the national, weekend press which sought to communicate to the public the benefits and safety of genetically engineered food. Two of the full page advertisements are reproduced in Figure 1.4(a) and (b).

Persuasion
To appeal or suggest to others the merits of a particular action, whilst accepting the possibility of refusal.

ACTIVITY 1.3

I would like you to scan the advertisements fairly quickly and then to re-read them with the following issues in mind.

1   The *tone* of the appeal – that is, the manner of presentation and the type of arguments advanced.

2   The *basis* of the appeal – namely, the way in which possible fears are allayed and the type of person chosen to endorse the use of biotechnology.

Write down your first thoughts on these different aspects and then give yourself some time to consider how anyone might be persuaded by the appeal. What is it exactly about the adverts which is designed to sway your opinion?

THIS STRAWBERRY TASTES JUST LIKE A STRAWBERRY.

FUNNY how we all suspect fruit and vegetables don't taste "as they used to." Year-round demand, forced ripening times and early harvesting are to blame.

Plant biotechnology offers the potential to produce crops that not only taste better, but are also healthier.

Monsanto is a leading biotechnology company. Our modified seeds are a new development of traditional cross-breeding, which has been employed for centuries. Each one is rigorously tested for safety and nutrition. The foods they produce have been approved by over 20 government regulatory agencies including those in the UK, Denmark, Switzerland and the Netherlands. Likely future offerings include potatoes that will absorb less oil when fried, corn and soybeans with an increased protein content, tomatoes with a fresher flavour and strawberries that retain their natural sweetness.

Monsanto believes you should be fully aware of the facts before making a purchase.

We support the efforts of retailers and others to provide you with labels about the use of biotechnology in food. We encourage you to look out for them.

For more information, ask for a leaflet at your local supermarket, call us free on 0800 092 0401, write to us or visit our website at www.monsanto.co.uk.

(STRAWBERRIES PRODUCED BY BIOTECHNOLOGY ARE NOT YET AVAILABLE IN THE UK)

MONSANTO
Food · Health · Hope

We urge you not just to accept our views. Please listen to other opinions. Call Iceland on 0990 133373 or visit their website at www.iceland.co.uk. Read what the Genetics Forum think at their website, www.geneticsforum.org.uk.

FIGURE 1.4(a)

FOOD BIOTECHNOLOGY.
LOOK WHO'S BEHIND IT.

"It is now time to recognise the huge benefits which genetically modified crops can bring to farmers, the food industry and consumers. After twenty years of research and scientific scrutiny, this versatile and safe technology is finally delivering the economic and environmental promise."

PROFESSOR MIKE WILSON, DEPUTY DIRECTOR OF THE SCOTTISH CROP RESEARCH INSTITUTE

"Biotechnology has the greatest potential to provide enormous opportunities for farmers along with great benefits for the environment and added choice for consumers. It could have an important role in maintaining the competitiveness of UK agriculture and horticulture and as such must be closely examined."

BEN GILL, PRESIDENT OF THE NATIONAL FARMERS' UNION

"To feed an increasing world population in the next millennium we are going to need genetically modified crops. To replace fossil fuels they will be vital. To help restore farming's place in society they will be instrumental. And provided every introduction is monitored and licensed by trustworthy experts I for one am looking forward to growing them."

DAVID RICHARDSON, FARMER AND CHAIRMAN OF LEAF
(LINKING ENVIRONMENT AND FARMING)

"Biotechnology will offer consumers, for the first time, a broader variety of foods that can provide a wide range of benefits not available before: enhanced vitamin content, lower fat, and better nutrition. Consumers will have more and better food choices than ever before."

PROFESSOR ALAN MALCOLM, CHIEF EXECUTIVE OF THE INSTITUTE OF BIOLOGY

MONSANTO
Food · Health · Hope

FIGURE I.4(b)

What struck me about the style of advertising was its measured tone: it is strictly 'factual', designed to engage with the seriousness of the issues involved and to demonstrate an awareness of alternative views on the subject of genetic engineering and food. There is no drama to the adverts, no obvious appeal to the emotions, nor any attempt to indoctrinate. But, of course, there are facts *and* facts – and in and of themselves facts do not make a compelling case for anything.

In this instance, the ability to persuade people to eat foods which they might otherwise not have done rests upon certain authoritative statements about the potential of biotechnology. Indeed, the voices are those of 'experts' or those who are well placed to be a source of 'authority' on the subject matter. Without the support of such voices, the factual basis of the arguments may possibly flounder and they would certainly lose much of their persuasive appeal. Equally, the promise of a better future (or better tasting strawberries at least) rests upon a privileged view of time which places the new seed technologies in a long line of traditional techniques. A certain degree of reassurance is also provided by the claim that rigorous testing procedures have been observed and government approvals have been forthcoming on that basis. Coupled with an insistence that you do not just take Monsanto's word for it and that you seek the opinions of others who oppose genetic interference with foods, the advertisements also work on the level of moral persuasion and 'balance'.

Naturally, the fact-based claims should be set alongside those of the scientific community at large, in which, as noted earlier, disagreement is rife. But the point of interest here, in relation to power, is not one of the type of evidence available; it is the way in which the company mobilizes 'tradition', 'fairness' and, above all, 'authority' to help carry the campaign.

Authority
Something that is claimed and, once recognized, serves as the basis by which others willingly comply.

The association of persuasion with **authority**, or rather with a number of authorities, turns on the distinction drawn by Anthony Giddens (1994) and before him by Hannah Arendt (1961) between those *in* authority and those who are *an* authority. The former refers to those individuals who occupy a particular institutional role, on the basis of which their authority and their ability to issue commands follows, as in the rank of the military or in a highly-orchestrated, rule-bound office administration. We shall return to this sense of authority in Section 4, when we discuss Max Weber's account of power and institutions. For the present, it is the latter sense of authority which interests us, where a person is *an* authority in a particular field because of their acknowledged expertise regardless of any institutional attachment.

In the past, and certainly up to the early post-Second World War years, the notion of an authority figure, someone whose judgement could be trusted, would probably have been given the benefit of doubt over the risks involved in some particular activity. Such authorities, in the public eye, dispensed a form of wisdom which could only be gained through access to specialized bodies of knowledge or expertise. The opinions of such figures would be deferred to on this basis and where disagreement among the experts arose, this was something that was best left to them to resolve, not something that the ordinary public could justifiably comment upon or debate.

Today, while the idea that there are authorities among us remains a persistent one, its strength has arguably diminished somewhat. According to Giddens, we no longer live in a period where trust in *an* authority is easily given or where faith in the 'experts' is willingly conceded. As a society, in the UK at least, it is said that we have become progressively disenchanted with those who claim to be grand 'experts', especially those that rely upon the authority of science to back up their claims (see also **Hinchliffe, 2000**). Faced with a diverse range of opinions on almost all matters, the public at large have learnt to be sceptical about 'absolute' truths. 'Grey areas' of knowledge have become the norm rather than the exception, and uncertainty a fact of life rather than an unusual state of affairs. Within such an unstable context, however, we still have to make up our own minds and so people continue to seek out expert advice and knowledge. But, says Giddens, the grounded scepticism of the contemporary age implies that our trust in experts is always revisable – and may be withdrawn as casually as it has been given.

If this is so, and we do not have to wholly endorse the argument to see its ramifications, then where does that leave Monsanto's advertising campaign?

Obviously its effect upon the consuming public is difficult to gauge, yet the mix of authority and persuasion, if Giddens' views are entertained, is not likely to be quite as potent as the company may have wished. Authoritative statements about the potential of biotechnology are likely to be treated with a degree of scepticism, as one view amongst others, rather than as a block of received wisdom. As if to confirm this point, one of the authorities cited in the advertisement later claimed that his comments had been clipped to convey a more positive assessment of the benefits of biotechnology than he had in fact voiced at the time (*The Guardian*, 12 September 1998) and indeed, in March 1999, the Advertising Standards Authority ruled that Monsanto's advertising campaign was at best confusing and at worst misleading over many of its 'scientific claims'. The 'facts', indeed, were not so compelling.

So in what other ways would it be possible for a large multinational to exercise power over consumers? If the intention is to have the public consume foods which they might otherwise not have done, how, for instance, could it achieve domination of the food market?

3.2 ... is it domination?

Domination, unlike authority or persuasion, is all about constraint and the removal of choice. Put another way, whereas authority is something that is conceded by those who recognize it, domination is imposed. There is a wide gulf between the experience of deferring to an authority, even if at a later

Domination
To impose upon or constrain the free choice of others despite possible resistance.

date we revise our opinions, and the experience of being subject to a dominant force. For domination to be secured in the world of food production and agrochemicals, a company like Monsanto, perhaps in league with others, would have to manoeuvre itself into a market position whereby consumers – indeed farmers, seed merchants, and the like – would have little or no choice other than to fall into line with the interests of the biotechnology multinationals. Despite public reluctance, all of those involved, from the farmers overseas down to those casually shopping at the local store, would be unable to escape the impact of the modified foods. In the language of 'hard' power, people would have no choice but to submit. We would end up eating the altered foods whether we want to or not.

Does this describe the current way in which power is being exercised in the food supply chain?

It is difficult to know for sure, of course, but for the situation to resemble anything like domination, the major players would have to hold a very firm grip on the food supply chain. For this to be the case, the big companies would at least have to gain control over crop research and production, as well as the sales and distribution networks. Seed companies, in particular, are pivotal to such an arrangement because they act as the distribution networks through which the big gene technology producers gain access to global markets. In this scenario, for instance, farmers in both the developed and the less developed world would be tied into the products of one company through restrictive corporate contracts or, worse still, bound in through seed stocks which left farmers no option but to purchase fresh seed from the same source year after year. This may sound a little far-fetched, but the genetic programming of plants to produce only sterile seeds, for example, is already a

FIGURE 1.5 Subject to domination? Farmers in the less developed world would have no choice but to purchase their seed from the big gene technology producers if only sterile seeds were available

proven possibility. If companies managed to manoeuvre themselves into near-monopoly positions across a range of geographical markets in such ways, the outcome would undoubtedly be close to that of global domination.

Before we go too far with this line of thought, however, we need to remember that other actors, such as supermarkets, are not part of this arrangement and that many governments, certainly across Europe, are less than willing to give automatic approval to genetically-altered foods. While it is true that the big biotechnology companies have, of late, initiated a spate of acquisitions which, in some cases, would leave them theoretically in control over almost everything from patenting down to the distribution of foods, the final say over market access rests with national governments. If governments say no to the widespread introduction of genetically-modified foods, then domination of the food markets by the big companies is simply out of the question.

Having said that, if a powerful American company like Monsanto had the backing of its government, then the USA – through the use of world trade rules – could insist that other countries, such as the UK, approve the company's genetic products. Is such an outcome likely or even feasible? It is impossible to be certain either way, but if it was remotely possible, then domination comes back into the picture as part of the arrangement of power.

This is perhaps as far as we can go with the analysis for the moment. Later, we will pick up on another way of thinking about domination. What we can do now is to draw together the observations on the different modes of power and consider how the latter relate to our elementary definition of power.

ACTIVITY 1.4

Think about this for a moment. Have the biotechnology companies exercised their power in such ways that people have eaten foods which they would otherwise not have done?

How would you begin to answer this question in the light of the various distinctions drawn between domination and authority, or between domination and persuasion, or between them all and manipulation?

Cast your eye, once again, over the key concepts in the margin of this section. Have the biotechnology giants exercised power in one or more of these ways at different times?

Well, I think it is clear that the companies have exercised a number of different *modes* of power in an attempt to shift public opinion and eating habits. Whether or not they have been effective remains to be seen however. If you have consumed genetically-modified tomato puree – probably the most visible GM product approved for sale in the UK – or certain modified vegetarian cheeses, without realizing their genetic history, then you can perhaps draw your own conclusions on this matter. But, more to the point, where in all this, you might ask, does that leave our *definition* of power?

For one thing, it shows that a two-line definition of power can take us only so far. Definitions have a role to play in grasping what may be the nub of something like power, its core characteristics so to speak, but they fall far short of an analysis of power relationships or an explanation of how they work. To capture the many-sided character of power – the different modes through which it is exercised, its various institutional sources, as well as its intended and unintended effects – we require a more extensive approach; one that examines power as a *theoretical* construction. On this basis, our aim should be not merely to grasp what power is but to understand how it has been *reasoned* theoretically.

SUMMARY

- The exercise of institutional power involves different modes, from coercion and manipulation to domination and authority. Each involves a distinct form of action and each corresponds to a quite different experience for those on the receiving end of a powerful action. When used in combination by large organizations, both public and private, they represent an institutional arrangement of power.

- Authority may be experienced in two ways: as someone *in* authority by virtue of their position in a particular institution or as *an* authority in some particular field of expertise. Arguably, trust in the latter form of authority has progressively been withdrawn in UK society to the point that uncertainty and grounded scepticism are the norm rather than the exception.

- To define something like power is to get a clear sense of it in one's head. Definitions represent a basic conceptual grasp of what it is that we are grappling to understand.

4 POWER AND INSTITUTIONAL LIFE

In what follows, we are going to look at two theoretical views of power which, whilst in agreement in many respects, differ primarily over how they see power exercised. The nub of the difference is over how they *reason* what happens when power is exercised. Why lay the stress upon reason? The emphasis is quite deliberate, for what is at issue here is the theoretical ability to explain many of the things that we consider important or interesting. To account for why something like the issue of GM foods took the course that it did in the UK involves reasoning towards a possible *explanation* of events.

And for this we need theory. Not, it should be stressed, as some kind of magical solution, but rather as a means to take us from the realm of definition to that of reasoned explanation.

The two accounts of power which we are about to examine can, at the risk of gross simplification, be whittled down to the following lines of reasoning.

- On the one side, power is reasoned as something which is held over others; a capacity possessed by certain individuals, groups or institutions who use it *directly* to secure their interests. Those that have power, those that are in authority, exercise it to get people to do things that they otherwise would not have done.

- On the other side, power is understood as an *indirect* affair, something which appears to be exercised behind our backs as it were, and is more about the way possibilities are closed down for us than any conscious decisions taken or executed by anyone.

Before we explore in greater depth the theoretical reasoning behind these two views of power – one based on the ideas of Max Weber, a German sociologist writing in the nineteenth/early twentieth century, the other based on the thinking of Michel Foucault, a French philosopher writing in the late twentieth century – let's first stand back and see where they broadly differ in their respective theorizations of power.

FIGURE 1.6   Max Weber (1864–1920)

FIGURE 1.7   Michel Foucault (1926–1984)

4.1 Two theories of power: Weber and Foucault

Perhaps a relatively accessible way to convey the differences in their understanding is to set them up in mock opposition. What is lost by this tactic, in terms of detail and points of agreement shared, hopefully is offset by the broad contrasts drawn.

If we start with the reasoning of power which has its roots in Max Weber's thinking, then:

- The issue of power and who gets to exercise it is quite straightforward. Either you have it or you do not have it; there is no 'half-way' house or in between situation. Those that hold power use it to further their objectives, despite possible resistance from others.

- In an institutional setting, be it a company office, a government department or educational institution, power is a *hierarchical, top-down affair.* The distribution of power is clear for all to see – some people give commands, while others obey – and each directs the actions of others further down the line of command.

- Obedience in such a setting, however, is not something that is automatically given. It has to be secured by those *in* authority and the basis for this varies depending upon whether we are talking about authority in the family, in the office, in the military, or wherever. Only where authority is seen as legitimate is the compliance of others forthcoming.

So, on this view, power is attributed to individuals or institutions who, in turn, exercise it in an intentional, rule-bound manner to dominate others. The only qualification we need to note at this stage is the somewhat precarious nature of authority: if it is not conceded, if it is not deemed legitimate, much of the rest of the institutional arrangement may become uncertain and rendered unstable.

By way of contrast, a different reasoning of power is evident in the work of Michel Foucault and like minded thinkers; one that does not locate *human agency* in quite the same way. On this account:

- Power is never in any one person's hands. On the contrary, power does not show itself in this manner. In fact it does not show itself in any obvious, willed manner, but rather as something which works its way into our imaginations and serves to constrain how we act.

- In the setting of a large workplace or government body, for example, the power of the institution does not pass from the top down; rather it *circulates* through their organizational practices. Such practices act like a 'grid', provoking and inciting certain courses of action and denying others.

- Compliance is not a straightforward matter, however, and turns on how far individuals *internalize* what is being laid down as 'obvious' or 'self evident'. Institutional power works best when all parties accept it willingly or, put another way, when they are persuaded to collude in their own subjugation.

This is perhaps a rather more difficult notion of power to grasp compared to the Weberian account, principally because it side-steps the question of 'who has power?' Once we remove the idea that power must be vested in someone – the company director or the institutional head, for example – it becomes that much harder to pin down exactly what power is or where it lies. Both views speak about *domination* as a mode of power and both ascribe a degree of uncertainty to the way in which it works. Where they differ, as noted earlier, is over how it is exercised: one sees it as a direct and visible relationship through which agents wield their legitimate powers, the other as an indirect practice internalized by agents who bring themselves into line.

Figure 1.8 gives you some idea of how the two contrasting theories relate back to the previous discussion of the different modes of power in play across institutions. You may find it useful to familiarize yourself with the broad differences in approach before moving on to a more detailed treatment of the reasoning behind each understanding.

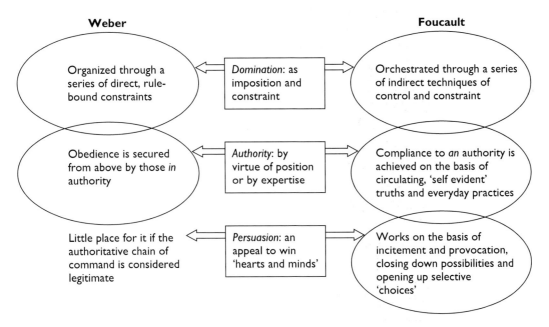

FIGURE 1.8   Two contrasting theories of power and their relationship to different modes of power

5 AUTHORITY FROM ABOVE ...

If we pick up again on the example of the big multinationals exercising their power across the markets of the world, it is quite possible that – if we were able to look inside their organizations – many of their actions would be explained along the lines of Weber's reasoning. For example, the access to specialized knowledges and expertise claimed by Monsanto, the attempt to conduct their campaign from an impersonal, scientific viewpoint, their busy acquisition programme to reap greater efficiencies of scale, and their reliance upon global rule-making bodies, all conceivably speak to one thing: the *rationalizing* work of power.

There is a kind of modern bureaucratic logic at work here which has more than a passing resemblance to Weber's account of institutional power. The example given may not quite match, say, the giant car plants in the early post-war period (which you will meet in Chapter 3) with their elaborately constructed hierarchies and formal system of rules and regulations, but from a contemporary angle there is sufficient in the example of Monsanto's actions to suggest the influence of a kind of rational, bureaucratic management, where discipline and control structures the everyday level of the organization. To what extent this is the case, we cannot be firm about, but for our purposes, Monsanto, as a modern multinational institution, like so many others, exhibits some of the key elements of **bureaucratic power**: namely, *a visible, hierarchical decision-making process based upon impersonal rules and specialist knowledges.*

Bureaucratic power
A form of institutional, rule-bound power based upon a clearly defined hierarchy of office.

Move beyond the boundaries of the organization to society at large and the same key elements can be seen to be at work in a variety of institutional settings. The methods may appear in a gentler guise, yet much of what we experience, for instance as a patient in a hospital ward or indeed as a student at a university, may be traced to the actions of those *in* authority deciding on an impersonal basis what is and what is not best for us. The decisions taken in such settings are not considered arbitrary or based upon personal whim. On the contrary, they are backed-up by qualifications, skills and expertise: a body of knowledge that in a rather matter-of-fact way announces solutions which, on the face of it, seem entirely rational and appropriate. Challenge the basis of such decisions and you run the risk of appearing 'irrational' or uninformed.

Much the same could be said for the plethora of rules and regulations which guide many of our actions in daily life. Public authorities lay down a series of prescriptions and prohibitions which serve to limit our actions – if you have a car, about how you should drive and where you can park, or if you have a job, the number of hours you can work and the appropriate health and safety measures you should take, and so forth. The rules are not set and maintained out of malicious intent, but rather because it would be irrational for those in authority to do otherwise. How else would it be possible to 'run' a modern, complex society or indeed a vast multinational concern without imposing some wide-ranging regulations and co-ordinating procedures. The 'rational' rules are there for a reason: to maintain order and to be seen to do so in an impartial and informed manner.

So how does the power of bureaucracy and officialdom work?

5.1 The powers of bureaucracy

The first point to note about the powers of decision making in a public or a private institution is that whoever holds them exercises them on *the basis of their position in the organization.* The administrators in a hospital or a private nursing agency, or their professional equivalent in a high technology or media company, occupy their decision-making roles by virtue of their technical expertise and competence to do the job. Authority comes with the position and with it the structural limits of what an official may dictate or prescribe. The discretion that an official may exercise, in relation to others who have to carry out their demands or in relation to those on the receiving end of such demands, is strictly circumscribed. Strict adherence to the rules and procedures embedded within an institution ensures that an element of personal favour or whim does not enter into the bureaucratic relationship.

This is a key aspect to consider, and it is worth reflecting upon in the context of modern-day management and the rule-bound nature of our lives. Imagine, for instance, being confronted by a government or a company official who constrained you to act in a way that you would otherwise not have done – but only because they took an instant dislike to your manner or appearance! Clearly you would be incensed and rightly so. The expectation is that such officials are only there to implement the rules and regulations in a just and equitable way, not on the basis of personal prejudice. And this, according to Weber, is the hallmark of modern bureaucratic authority: divorced from personal ties and influence, the position of the administrator is one of 'duty', a vocation which finds its true expression in the *impersonal official.*

This takes us to the second point to note about the way in which bureaucratic structures work. The separation of the person from the position of power is not something that is simply preferred, it is 'built in' so to speak to the machinery of bureaucratic power. In Weber's view, the organizational strength of bureaucratic forms of administration lay with its clearly established *hierarchical structures of supervision and control.* Each person occupies a certain position in the vertical chain of command and each has a specific task to perform which can only be executed on the basis of orders received from above. Thus while officials exercise power in a downwards fashion, they may only do so within the strict limits of the rules laid down by the bureaucratic structure.

Put another way: individual *agents* exercise power but only within the possibilities generated by the institutional *structure.*

Or, in Weber's words:

> The individual bureaucrat cannot squirm out of the apparatus into which he has been harnessed. ... In the great majority of cases he is only a small cog in a ceaselessly moving mechanism which prescribes to him an essentially fixed route of march. The official is entrusted with specialized tasks, and normally the mechanism cannot be put into motion or arrested by him, but only from the very top.
>
> (Weber, 1978, pp.987–8)

From your own understanding about how large organizations work this statement may seem somewhat exaggerated in respect of how little leeway is available to individual agents within the structure. None the less, what is readily apparent is the highly visible nature of the exercise of that power, in

FIGURE 1.9 Cogs in a machine? Bureaucracy prescribes the limits of individual action and discretion

particular the written and codified body of rules which make the assessment and calculation of individuals a transparent process. More to the point, clearly-structured hierarchies also make it possible to locate responsibility for decisions taken and to administer the appropriate redress should such decisions be considered partial or prejudicial.

A third and final point to consider about the nature of bureaucratic power is the extent of its operation. A hierarchy of control is a useful means of co-ordinating the complex and sometimes elaborate tasks of running a business or administering public policy, but it is also a means of *extending power through delegation*. Think, for example, of what it must take to effectively run one of the big biotechnology corporations across far flung national borders or what is involved when a central government agency sets out to administer a uniform standard of social welfare across the country as a whole. The possibility of micro-management from the top is an unrealistic option, despite the existence of a common set of rules and regulations. In principle, the answer is to delegate authority, not power.

The delegation of authority down through a tiered hierarchy of position-holders has the effect of expanding the scope of the powers held by those at the top, be it the company head, public chief executive, or whoever sits at the apex of the institution. Power in this sense can be seen to radiate out from an identifiable centre, even though at the furthest extent of the hierarchy it is experienced as a layer of authority. Those officials engaged directly in the business of executing commands or in, say, the assessment of welfare benefits at the local level should, in theory, operate with the minimum of discretion to realize the instructions of the centre.

In practice, however, it does not seem to work quite like that.

5.2 Front-line authority

Authority on the front line, so to speak, involves administrators and officials in direct contact with people on the receiving end of bureaucratic edicts. In the public sphere, for instance, delegated officials are expected to apply the rules to everyone on the same basis, regardless of background or lifestyle. The rules are general and universal in application. However, people are neither. Their particular circumstances vary, their needs are often specific not general, and in many cases the universal models provide an awkward fit. In such circumstances, front-line officials may find themselves unable to simply transmit the instructions from the centre and reluctantly forced to draw upon their own discretion in making judgements. In doing so, however, they run the risk of *losing* their authority.

Recall the distinction drawn earlier between those *in* authority and those who are *an* authority, where the former refers to those who hold authority by virtue of their institutional position rather than any specialized knowledge claim. An expression of duty, rather than a specific expertise, justifies the position of bureaucratic authority and the legitimacy of such a position is conceded by others for as long as that duty is impartially performed. Overstep that authority by, for instance, interpreting the rules to fit the situation and the legitimacy may simply evaporate overnight. (The legitimate basis of patriarchal authority is rather different and is outlined in Box 1.2.)

BOX I.2   Patriarchal authority

In *Economy and Society*, Weber distinguishes between patriarchal domination and bureaucratic domination principally on two grounds. In the first place, where bureaucratic domination is based on an official's commitment to a sense of impersonal duty, the former relies upon the *personal authority* of the master, the *male* head of the household. The head of the family is said to possess authority on the basis of personal relations which are considered natural and enduring. This leads to a second difference between the two forms of domination, in so far as patriarchal authority is rooted in *tradition* passed down from one generation to the next and bolstered by filial bonds. The obedience of the wife and children to the male head is said to derive from dutiful conduct rather than from a set of rules established on an abstract, rational basis.

This rather definitional account of patriarchal authority is developed in a more 'everyday' setting in Chapter 2.

Barry Barnes (1988) draws attention to this possible scenario as part and parcel of the nature of power. Perfect authority, he argues, is impossible to maintain; those further down the chain of command will inevitably exercise their delegated powers with discretion simply because there is no other way. The net result is a more provisional sense of institutional power, where control must necessarily be incomplete and the ability to order people's lives always imperfect. Weber's notion of the individual bureaucrat as a mere cog in the administrative machine is, in that sense, an overstatement. Outside of the public sphere, in the lower ranks of the big commercial companies for instance, authority at the office level must also involve a reasonable element of discretion and an openness to reinterpretation.

It is perhaps through the example of welfare state agencies, however, that we can readily see how the *dispersal* of power has led to a less stable arrangement of institutional power than is commonly thought. For theorists like Barnes, the delegation and the dispersion of power are one and the same thing: once authority is devolved, delegates are empowered and an independent use made of that power. Similarly, for John Clarke and Janet Newman (1997), the delegation of state power in the field of social welfare has led to its dispersal through a variety of quasi-autonomous agencies in the private, voluntary and informal sectors. What is involved here, according to Clarke and Newman, is an identifiable shift in the way that power descends through the social welfare hierarchy. In place of direct structures of supervision and control, the new front-line agencies – the trusts, voluntary organizations, subcontracted firms and the like – are held in check by monitoring and contractual arrangements exercised by the centre. You will consider these new arrangements in greater depth in Chapter 4, but for now all that you need note is that, as before, such delegation and dispersal of authority may enhance the power of the centre, yet it also opens up a greater possibility for the distortion and ambiguity of decision-making practices.

On this view, what was perhaps always a less complete, less perfect line of public command than was believed to be the case, is now an inherently unstable structure of power with a greater number of outside interests to accommodate. The mix of public, private and voluntary agencies involved in the exercise of discretionary judgement and the interpretation of the rules does in that sense draw attention to the fact that the state's ordering of lives has never been a straightforward top-down affair of domination.

And this view, in many ways, chimes with the starting point of Foucault's reasoning on power, although as we shall see he takes us in a quite different direction from that of bureaucracy, dispersed or otherwise.

SUMMARY

- The machinery of power, for Weber, is exercised by *agents* who do so within the limits of their position in the institutional *structure*. Those *in* authority attempt to order people's lives through the 'rational' character of rules and regulations which are more or less impartial, more or less universal.

- *Organized domination* orchestrated by those in authority encapsulates the way in which Weber characterizes institutional power. The removal of choice and the imposition of constraint lie at the heart of the process.

- Institutional power, however, is extended through delegation, with the attendant risk of losing authority or destabilizing the decision-making process. It is therefore less a monolithic force, ordering all in its path, and rather more a provisional set of arrangements, the effectiveness of which cannot be guaranteed in advance.

6 ... OR GOVERNING THE SELF?

If the orchestration of people's lives on a day-to-day basis is not quite as top-down as modern bureaucratic logic would suggest, it remains valid to think that someone is still directly running things 'up there'. Someone or some group in a position of authority, whether at the top of a big government department or indeed on the board of a multinational company like Monsanto, continues to hold the reins of power. For Foucault this is simply not the case. We are so used to thinking about power as an overt and identifiable force that it is hard to conceive of power as a guiding force which does not show itself in an obvious manner. Your responses to Activity 1.1 may have been revealing in this respect. Certainly, for my part, when on the receiving end of what appears to be an arbitrary decision or sanctions on my behaviour my immediate reaction is to think about who, *which* agency, lies behind it – which faceless bureaucrat or figure of authority? My simple error, if Foucault is to be believed, was to think about power as an identifiable force in the first place. It, we are told, possesses an altogether more anonymous quality.

It is hard to know exactly what to make of this. But rather than consider ordered lives as the product of consciously elaborate rules and prohibitions, we should perhaps be turning our thoughts as to why, when faced with all kinds of possibilities, we act in the ways that we do and not otherwise. When no direct constraints are placed upon us, why do we limit our behaviour in certain ways?

Take the example of the hospital patient or the university student again. Why is it that, when little or nothing has been said, do people exercise self-restraint in their role as patient or as student? What prevents most of us from immediately challenging the basis of the doctor's or the lecturer's authority? In part, it is because, as Charles Taylor (1986, p.184) has indicated, both sets of relationships rest upon the presumption that one side 'knows and that the other has an overwhelming interest in taking advice'. It goes without saying that the doctor, for instance, does not set out to undermine or hurt the patient and that the patient willingly places themselves in a subordinate position in the hope of a cure or a better life. The relationship, as we have seen, may be one of authority, but it is entered into willingly on the part of the patient.

The pressing issue here for Foucault is that we take it upon ourselves to regulate our own conduct. Even though we are free to act in all kinds of ways, we choose to constrain our behaviour. And the reason why is that we know what is expected of us. It is this framework of expectations, presumptions, ideas which 'go without saying', that ensure compliance. At the broadest level, for instance, government bodies work within such frameworks, diffusing standards and exhorting ideals which are hard to find fault with. After all, in the arena of social welfare, as we shall see in Chapter 4, who does not broadly agree that 'taking responsibility for our own lives' or responding to a 'hand up' not a 'hand out' from government is the right thing to do?

Far from signifying total domination, such inducements seem intuitively obvious. Indeed, normal. It is in this sense that power works as an *anonymous force, provoking free agents to act in ways that make it difficult for them to do otherwise.*

Let's turn to this understanding now to see how such a force operates.

6.1 The powers of provocation

The starting point of Foucault's account of power is something of a paradox: rather than power curtailing the freedom of individuals, it works on the basis that *we are free to govern ourselves*. Quite simply, in the absence of any sanctions we still opt to restrain our behaviour. There is no overarching structure of power which brings us into line, only the acknowledgement that people operate within a framework of choices which are themselves subject to influence and direction. In that sense, power is a positive rather than a negative force; it enables people to fashion their own lives rather than repress any such possibility. The provocative advertising campaign launched by Monsanto can perhaps be read in this light. Ask people to make up their own minds about GM foods, to think through the possibilities for themselves, and they may well develop the sort of 'responsible awareness' that the company clearly believes is so oddly lacking in the UK.

As we have seen, however, it would be a mistake to take this line of reasoning too far. Although power is conceived as an open-ended affair, it is none the less viewed by Foucault as a stabilizing force which leaves little room for manoeuvre. The stress is upon the ways in which institutions – whether it be a private company, welfare organization or indeed a household – close down possibilities rather than proliferate them. This, then, is not about provocation for its own sake, but rather **provocation** as a means to order people's lives in particular ways.

Provocation
To incite or induce a certain course of action.

In a wider organizational setting or in the family household, for instance, on this view it is not so much the rules which govern daily life, as it is the ideas and practices which are suggestive of what is appropriate and what is inappropriate behaviour in these settings. This, in many ways, is the very ground of advertising, where companies like the biotechnology giants attempt to mould our reactions to the new food technologies. Equally, in relation to the family, we may possibly think, for example, that in challenging the authority of the father-figure in the family setting we gain a certain freedom, yet in Foucault's reasoning we are just as likely to remain *dominated* by ideas and images associated with 'normal' family life and its relationships of parenting, and the like. In short, as will become clear in Chapter 2, we remain 'trapped' in our thinking about the nature of 'proper' families, even though we may have cast aside the shackles of patriarchal authority and the traditions of male order.

There is a further dimension to all this that we need to bear in mind, however, which is that such ideas are not simply free-floating. Rather, they are rooted in *institutional ways of doing things, techniques and practices which more or less call forth appropriate behaviour.* The number of possible settings appears vast, ranging from the hospital ward, the lecture theatre, or the family home as we have seen to the office floor, the prison yard, or the local welfare benefits office. It is in this sense that power is dispersed for Foucault. In contrast to Weber, there is no aloof centre from which power descends, only a certain 'roundaboutness' to power in which each institutional setting evokes a particular form of control – be it through the physical layout of a setting, the ordering of routine activities or simply through the evocation of particular 'ways to be', or indeed a combination of these techniques (Foucault, 1980).

Perhaps one of the more obvious illustrations of how such techniques can work is that of the telephone 'call centre', where a workforce, the majority of which are likely to be women, are subject to a range of *indirect* techniques of control (see **Mackintosh and Mooney, 2000**). Call centres are an expanding sector of employment across Europe as whole, but they also introduce new organizational and managerial techniques which illustrate something of the 'roundaboutness' to power in a contemporary work setting.

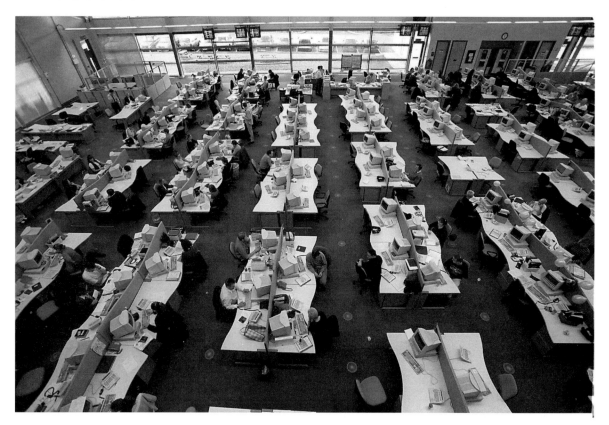

FIGURE 1.10   The setting of a telephone call centre: a provocation of power?

Look closely at the work arrangements in Figure 1.10, in particular the physical layout, and try to imagine what it would be like to work in such an institution. Some of you may know already what it is like: the precise ordering of routine activities, the overarching role of the supervisor, and the close monitoring of performance and behaviour (Richardson and Marshall, 1999). In one sense, all very Weberian – top-down and closely regulated. But there is also an *indirect* side to the organizational arrangements of control in call centres. Bearing in mind the issue of surveillance, can you think what those indirect techniques might be?

In all likelihood there will be some concern over the politeness of the telephone operators and an inducement to self-monitor performance. Yet employees are probably under no illusion that these aspects of their performance are subject to some form of direct regulation. However, what is striking about the nature of the monitoring is that, for much if not all of the time, those taking the calls are aware that their every action may be under observation, yet are themselves unable to know whether this is actually happening. The net effect of this is for employees to discipline their own behaviour so that it conforms to the expectations of those *who may be* monitoring their performance.

This comes close to the heart of Foucault's reasoning on power which, as noted in the example of the hospital patient earlier, revolves around *indirect techniques of self regulation* which induce appropriate forms of behaviour. It matters less in this context whether the monitoring techniques work effectively or not, only that those answering the phones accept the 'truth' of the arrangements. At the point where the workforce *imagine* that they are being monitored, this is precisely the moment when, in Foucault's sense, they bring themselves into line and assume the role that has been indirectly carved out for them.

Expressed in a different manner: power works on and through *agents* in ways which structurally *limit* what they might otherwise have done.

Whether or not such arrangements are as effective as claimed, though, is a more open question.

6.2 Hit-or-miss domination

In Section 3, I spoke about domination as a mode of power in which people have little or no choice but to fall into line with the constraints imposed upon them. The twist that Foucault gives to this understanding of domination is that it is through people 'working' on their *own* conduct that they bring themselves to order. At the level of the ongoing running of institutions on a day-to-day basis, individuals *internalize* what is expected of them because it seems the right and proper thing to do. If this sounds less than total domination, that is because at best it represents a modest form of domination.

Governing the self, as we have seen, operates within a framework of choices which must allow for the possibility that individuals resist or even reject the organized practices which, for example, circulate through call centres or the suggestive influences and pressures of an advertising campaign. In the absence of coercion or the likelihood that the new 'open' styles of communication engage in deliberate attempts to manipulate or dupe people, the relationship of power is probably closer to that described by Foucault as one of 'permanent provocation' (Foucault, 1986). By this, Foucault seems to be edging towards a scenario of power in which each side circles the other, vying for position in the hope of influencing the outcome of the game-play.

In many ways, this kind of scenario strikes a chord with that described at the end of Section 5. While the issue here is not one of delegated powers or the exercise of discretionary judgement, the uncertain outcome of an altogether more modest form of power highlights the provisional nature of institutional power. That such institutional arrangements may long have been characterized by instability and uncertainty may only now be in the process of being fully recognized.

SUMMARY

- Institutional power, for Foucault, should be understood as something which works on and through *agents* who, in turn, exercise self discipline and restraint.

- Order in contemporary society is thus achieved largely by individuals governing themselves according to what they imagine to be the most appropriate or acceptable forms of behaviour.

- *Domination*, for Foucault as well as for Weber, characterizes the outcome of institutional power, but for Foucault it is a state of affairs brought about by indirect techniques and received 'truths', rather than by organized, rule-bound practices.

- Institutional power is not so much above us, as all around us. It touches everyone in ways which are hard to pin-point, yet difficult to rebut. It is, none the less, far from an overarching scheme and closer to a provocative game-play in which all outcomes are unstable and open to revision.

7 THE POWER OF THEORY

We have come a long way in our examination of Weber's and Foucault's accounts of power. In particular, we have moved from a brief definition of power to a more elaborate theoretical construction of power, or rather constructions, and in the process we have spelt out some of the key aspects of power which characterize Weber and Foucault's thinking on this issue. But, I can almost hear you say, where does that leave us in terms of *explaining* anything? Where is the pay-off for theory in all this?

To answer this question, we need to go back to the beginning of Section 4 and to the stress placed on the powers of theoretical reasoning. Theorizing, as something which social scientists do, involves a large degree of thinking: that is, thinking carefully about the concepts they are working with, whether it be domination, authority, compliance, or whatever, and thinking systematically about the connections between them and how they may be put together to actually explain something.

When expressed like this, the practice of theorizing sounds relatively easy. Why all the concern then? Well, needless to say, it is not as easy as it sounds. The difficulty is that theorizing – as a practice – is not always so straightforward to follow. As a mode of thinking, the threads of theory often have to be unpicked before they become apparent. We can start by looking at the *questions* that both Weber and Foucault ask about power before moving on to see how such questions shape their explanation of events, in this case, by returning to the GM foods controversy and the agencies involved.

If you turn your attention to Table 1.1, to the left-hand column, under the heading of *questions*, you will see the sharp contrast in the questions asked by Weber and Foucault about power and its institutional arrangements. In fact, these questions were tentatively raised in Section 4.1, with specific mention made of the point that Foucault side-steps the question of 'who has power?' This implies, of course, that for Weber the questions of 'who' is in authority?, and 'who' controls the power apparatus?, are the crucial ones. Weber wants to know 'where' power lies and 'who' possesses it and, by dint of that, 'who' does not. In contrast, Foucault is more concerned with 'how' questions: 'how' is power exercised?, and 'how' does it circulate?

TABLE 1.1 The threads of theory

	Questions	Theoretical claims	Evidence
Weber	• Who holds power? • Who controls the rule-making machinery?	• Domination by authority involves the imposition of rule-bound constraints on the conduct of others. • Bureaucratic power is a rational, top-down affair with clearly-defined lines of authority and delegation.	The types of evidence sought include the *visible* actions of governing bodies, in particular: • the forms of expertise and institutional authority drawn upon, and • the rule-making process.
Foucault	• How is power exercised? • How does power circulate?	• Domination works on the basis of self-restraint rather than external constraint. People bring themselves to order. • Power is provocative, it is brought to bear on people's actions, closing down rather than opening up possibilities.	The types of evidence sought in this instance are more *elusive*, in particular: • the indirect techniques and practices which routinely 'govern' our lives, and • the ideas and accepted 'truths' which influence our behaviour.

In considering the GM controversy, then, Weber's thinking diverts our attention to those who are in control. It directs our attention, among others, to those at the top of the big biotechnology companies, to the 'decision makers', the managing directors, chief executives, and high-ranking officials who orchestrated the company's strategy and marshalled 'expert' opinion. Notice, as well, that Weber's questions also give us a *starting point* to the theoretical inquiry, namely an organized hierarchy with a clear sense of who is at the top and who is at the bottom.

Foucault's thinking, however, takes us in a different direction altogether. It directs us to the circulation of power, that is, to a net-like organization which, in the GM context, would encompass relationships within and between Monsanto, government authorities, environmental scientists, seed companies, farmers, consumers and many more. Foucault's questions would, for example, draw our attention to an array of techniques ranging from the persuasive tactics of the advertising campaign and the organization of field trials to the provocative game-play between all those involved, especially the line-up between corporate and environmental interests.

What, then, is happening here in terms of the process of theory construction?

Well, in both instances the questions posed by the thinking of Weber and Foucault move us on – in different ways – to a new set of relationships. Their focus is selective, in so far as they home in on certain relationships at the expense of others. In doing so, however, they also suggest relationships between things that as yet we may not have noticed and which lend weight to a small number of concepts. There is, for want of a better way of describing it, a kind of bundling process at work in which certain concepts are brought together by the two theorists in a manner that is suggestive of particular *lines of reasoning* which, in turn, lead to definite *theoretical claims*.

These claims are set out in the middle column of Table 1.1. To be honest, if you followed the theoretical discussions in Sections 5 and 6, there is little that will surprise you. When considered against the backdrop of the GM controversy, however, it is revealing just how far the contrasting logics of the two positions takes us substantively. In Weber's account, for instance, the concepts of domination and authority are linked to a rule-bound hierarchy in a way which suggests that any possible resolution to the issue is likely to turn on the legitimacy of both the rules and the 'voices' of authority. Unless, that is, domination is achieved through the elimination of consumer choice by the actions of the big bureaucratic corporations and global rule-making bodies. Whereas, for Foucault, domination holds out a rather different scenario, whereby the effectiveness of any campaign in favour of GM foods is likely to be judged by how far people internalize genetic modifications as an 'acceptable' technology. When we can no longer recall what all the fuss was about, then, and only then, can the techniques of inducement be deemed successful.

Of course, the above scenarios are merely speculative and, without recourse to *evidence*, the robustness of either theory is an unknown quantity.

Suffice to say that, for now, it is useful to note that theories also guide the *selection* of evidence and influence what *counts* as evidence. So, if we were to extend Weber's theory of power into the realm of GM foodstuffs, environmental agencies and institutional machinations, the kind of evidence that is likely to be mulled over would concern, for example, the questionable authority of high-ranking officials in both corporate and government bodies, the acquisitions programme carried out by the big multinationals to extend their geographical reach, the global rule-making exercised by the World Trade Organization and their efforts to constrain, and so forth, which, all in all, may have led the public to eat certain foods which they would otherwise not have done.

ACTIVITY 1.6

Now do the same kind of exercise for Foucault and the types of evidence likely to be valued. Look back to Table 1.1, to the final column, and read the entry on evidence under Foucault. What kind of indirect techniques in circulation or 'truths' about the appropriateness of gene technologies would one be looking for?

Don't consider making a list at this point, rather treat the exercise as a way of re-capping some of what has been said earlier about Foucault in this chapter. You'll return to both Foucault and Weber on power shortly.

8 CONCLUSION: ORDERED LIVES

The purpose of the chapter, as stated at the outset, was to explore the nature of power and how different institutional arrangements of power are put together, theorized, and drawn upon to explain the various changes that we have witnessed in, for instance, the make-up of the family, the significance of work today, and our relationship to the state and its welfare provisions. Each of these institutional settings is explored in depth in the chapters that follow.

Having said that, the chapters add up to more than a description of change at the beginning of the twenty-first century. For one of the things that will be asked of you is to grasp both continuity *and* change in these key areas of social life. Put another way, you will be asked to think about how an institution like the family is in many ways an ordered, stable fixture of people's lives yet, at one and the same time, a source of instability for many as it changes in both form and substance. Hopefully, what this chapter provides are some of the tools, as it were, that will assist you in that analysis.

As an introduction to power and order, the preceding pages have introduced you to the various ways and means by which power seeks to achieve order, from the more obvious rule-bound procedures, official regulations and expert knowledges to the indirect techniques, organized routines, and suggestive ideas which may serve to limit our imaginations. Depending upon the setting and the institution in question, different modes of power will underpin the various arrangements. Also, as you will have gleaned from the last section, different theories will be selective about the ways and means that they consider important.

But perhaps one of the most significant messages that you should take from this chapter is the less than certain outcome to many of the institutional arrangements of power. The stress in this book placed upon the process of *ordering*, rather than simply order, is suggestive of the more provisional nature of social institutions in contemporary society. For confirmation of that, you need look no further than the subject of the next chapter, the changing institution of the family.

REFERENCES

Arendt, H. (1961) *Between Past and Future: Six Exercises in Political Thought*, London, Faber and Faber.

Barnes, B. (1988) *The Nature of Power*, Cambridge, Polity.

Clarke, J. and Newman, J. (1997) *The Managerial State: Power, Politics and Ideology in the Remaking of Social Welfare*, London, Sage.

Foucault, M. (1980) *Power/Knowledge: Selected Interviews and Other Writings 1972–1977*, Brighton, Harvester.

Foucault, M. (1986) 'The subject and power' in Dreyfus, H.L. and Rabinow, P. (eds) *Michel Foucault: Beyond Structuralism and Hermeneutics*, Brighton, Harvester.

Giddens, A. (1994) 'Living in a post-traditional society' in Beck, U., Giddens, A. and Lash, S. (eds) *Reflexive Modernization: Politics, Tradition and Aesthetics in the Modern Social Order*, Cambridge, Polity.

Hinchliffe, S. (2000) 'Living with risk: the unnatural geography of environmental crises' in Hinchliffe, S. and Woodward, K. (eds).

Hinchliffe, S. and Woodward, K. (eds) (2000) *The Natural and the Social: Uncertainty, Risk, Change*, London, Routledge/The Open University.

Mackintosh, M. and Mooney, G. (2000) 'Identity, inequality and social class' in Woodward, K. (ed.).

Richardson, R. and Marshall, N. (1999) 'Teleservices, call centres and urban and regional development', *Services Industries Journal*, vol.19, no.1, pp.96–116.

Taylor, C. (1986) 'Foucault on freedom and truth' in Hoy, D.C. (ed.) *Foucault: A Critical Reader*, Oxford, Blackwell.

Weber, M. (1978) *Economy and Society, Vols 1 and 2* (Edited by Roth, G. and Wittich, C.), Berkeley, University of California Press.

Woodward, K. (2000) 'Questions of identity' in Woodward, K. (ed.).

Woodward, K. (ed.) (2000) *Questioning Identity: Gender, Class, Nation*, London, Routledge/The Open University.

FURTHER READING

On the different modes that power can take, an accessible little text, that has its roots in the difference between violence and power, is Hannah Arendt's *On Violence* (1970, Harcourt Brace). Section II is the most relevant. The ideas expressed there hold considerable relevance today.

A more recent and illuminating account of institutional power is Davina Cooper's *Governing Out of Order: Space, Law and the Politics of Belonging* (1998, River Oram Press). Influenced by Foucault's writings but also Weberian ideas, this book is exceptional for its attempt to explore both direct and indirect forms of power in contemporary contexts.

Family: from tradition to diversity?

Norma Sherratt and Gordon Hughes

chapter 2

Contents

1 INTRODUCTION

It is difficult to read a newspaper, listen to the radio or watch TV today without coming across urgent voices expressing concern over what's happening to the 'traditional family' in contemporary UK society. The focus may be the growing numbers of lone parents, 'artificial' or 'unnatural' examples of reproduction opened up by technology, debates about same-sex marriages, or increasing rates of divorce and separation. We may ask ourselves 'what's going on here?' As with the topic of crime (Mooney *et al.*, 2000), popular discussions of the family seem to be characterized by both fear and fascination. There is the widespread fear that changes in family lives are leading to greater uncertainties and private troubles in people's lives. We have, then, a picture of a loss of order and certainty. At the same time, we seem to be fascinated by the departure from the 'old' ways of the traditional, father-dominated family in which simple assumptions of what was and wasn't 'natural' predominated. There appears to be a greater diversity today in the ways that our intimate and domestic living arrangements are organized. But does such diversity represent a loss of order or rather new ways of ordering our family lives with new constraints and pressures?

This chapter seeks to explore these debates chiefly through the theories of power which John Allen introduced in Chapter 1. Are changes in family life, for example, most usefully understood as a move away from traditional patriarchal authority? Or does the emphasis on negotiation and equality in some families today only serve as a disguise for a continuation of men's power? Is it the case that how we order our lives in families has always been dependent on *ideas* of what constitutes 'normal' and 'natural' family life? Furthermore, though these ideas may change (we no longer, for example, see husband as breadwinner/wife as homemaker as the obvious way of dividing responsibilities in the family), they do not do so to the extent that we start to question 'the family' itself. Does the persistence of appeals to something called a 'family' across quite diverse domestic arrangements mean that we can't, in Foucault's terms, 'imagine' ordering our lives outside the norm of the family? Seen this way, diversity is not synonymous with a breakdown of an institutional order but rather represents a new ordering. Such theorizing also helps us explore the moral and political debates around the 'decline of the family' thesis. In this chapter we focus on two **political ideologies** as opposing ways of thinking – conservatism and feminism – which have informed this debate both within and beyond the social sciences.

Political ideologies
A cluster of ideas that both describe societies, propose moral and practical alternatives, and are aligned with political movements.

In Section 2 we compare the idea of a 'golden age' of the traditional nuclear family with the changing institutional arrangements of family life today, drawing on both qualitative and quantitative data. Section 3 then compares two sides of the debate about the 'decline of the family': the political ideologies of conservatism and feminism. Section 4 explores the changing ways in which power relations order family lives past and present, drawing explicitly on the theories of Weber and Foucault. Finally, in Section 5, we consider the family as a site where, over recent decades, social and economic inequalities have been deepening and reproducing themselves.

2 TRANSFORMATIONS IN FAMILY LIVING: THE 'GOLDEN AGE' AND BEYOND

Most people in the UK have been born into and brought up in families. In turn most of us will have children and form families. It is not surprising that the family often appears to be the most natural, normal, taken-for-granted feature of the ordering of everyday life. Yet it is difficult to talk of the family today without reference to the decline of the 'traditional' family (however defined) and the break with the past, seemingly ushered in by new living arrangements at the end of the twentieth century. This most intimate of all social institutions is thus now often spoken of in terms of change and uncertainty. But what do we know of family life in the 1950s? What do we mean by the 'traditional' family? Did family life for most people correspond to the idealized image which was promoted in the 1950s (just as it has become our reference point today)? And what evidence do we have of changes to family life in the following decades? These are the questions we as social scientists need to ask as we explore the complex processes of social change by both challenging common-sense 'certainties' and by careful interrogation of claims, evidence and theories.

2.1 Happy families of the 'golden age'?

Let's begin by outlining the main features of the often idealized 1950s family. In this period of major social reconstruction following the disruptions and uncertainties resulting from the experience of the Second World War, great significance was given by policy makers and moral commentators to the role of the **nuclear family** in forging and ordering the 'healthy' society. The dominant expectations of family life in the immediate post-war period in the UK were that marriages were for life, married men were responsible for their partners and children, men's place was in the world of the external labour market, and women's place was in the home caring for home and children.

Nuclear family
A social unit consisting of wife, husband and dependent children.

ACTIVITY 2.1

The two photographs in Figure 2.1 are representations of this kind of family life. Note for yourselves now what they tell us about being a father, mother or child in the 1950s. You will find the differences between the two photos as interesting as the similarities.

(a)

(b)

FIGURE 2.1   Family life in the 1950s: (a) at home and (b) on a holiday weekend

COMMENT

Analysis of any image or set of images involves exploring the social context in which they are produced and consumed (termed a horizontal analysis) as well as their informational content (termed a vertical analysis). It is undoubtedly of importance to recognize that these were photos taken in the 1950s as part of a celebration of a new family togetherness. As such, the information they give us cannot be complete. But a careful reading can, none the less, take us a long way in exploring not only the expectations but also the lived experiences of mothers, fathers and children at that time.

The two photos are undoubtedly similar. Both are unmistakable images of the nuclear family of the 1950s, not of family life even a generation earlier, or of family life today. The father is not represented as a traditional patriarch, distant and powerful, sitting at the head of the table, but as part of an informal domestic family grouping. In each image the family is spending time together, in the same room (arguably one aspect which marks it off as not belonging to the start of the twenty-first century). And children's play is an important part of that time (you may have noticed the way in Figure 2.1(b) it is the children who are central to the picture, facing the camera). These are images of harmonious companionship and shared enjoyment of family life based on the ideal of 'companionate marriage' (Finch and Summerfield, 1991).

And if we compare the two images we can get closer to understanding the basis of this companionship and sharing. The father in photo (b) is dressed informally, sitting at the table with the mother and children, only because it is the weekend when he does not go *out* to work. In photo (a) they are certainly all together; yet the father is also separate. The domesticity of the woman knitting and the children playing is contrasted with the man who has perhaps just come in from work – dressed for and involved still in activities which set him apart. His position at the centre of the group and at the same time close to the window could signify him as the provider and the link to the outside world.

So in this idealized picture mother and father are both seen as contributing to family life. But their **roles** and contributions are seen as essentially different. With marriage portrayed as teamwork, the role of the mother was exclusively bound up with serving the needs of both the husband/father and the children in the home. Child rearing and paid work were assumed to be mutually exclusive. And the husband/father was seen as the 'breadwinner' responsible for supporting the family through his paid work outside the home.

Optimism of this kind about the post-war nuclear family was also reflected in some of the most influential sociological studies of this period. Young and Willmott's (1957) and Willmott and Young's (1960) studies of working-class family life, for example, pointed to the great improvements in the conditions

Roles
The social expectations and prescriptions which go with different occupations and social positions.

of working-class households when compared with earlier generations of studies. They also celebrated the new 'companionate marriage':

> There is a new kind of companionship, reflecting the rise in status of the young wife and children which is one of the great transformations of our time. There is now a new approach to equality between the sexes and though each has its peculiar role, its boundaries are no longer so rigidly defined, nor performed without consultation ... man and wife are partners.
>
> (Young and Willmott, 1957)

But did family life for most people correspond to these images and accounts? This idealization of the nuclear family in the 1950s – conjuring up a 'golden age' – certainly tells us nothing about the dark secrets of abuse and hidden violence which later researchers have revealed. But we also need to be aware of the implications for people's lives in all families of the unequal power relations which are fundamental to this kind of nuclear family. The man, as breadwinner and also point of reference between the family and the state, was in authority (see Chapter 1) as the 'head of household'. And although, as you will see in Section 3, there is a disagreement about the significance and consequences of these inequalities, they were undoubtedly part of family life at that time. Women and children were dependent on men, the lines of power were seemingly quite clear cut.

2.2 Contemporary trends in family life

In the second half of this section we will be looking chiefly at statistical material which plots the main trends in families and households in post-war UK society in order to get as comprehensive a picture as possible of what family life is like today as compared to previous decades. Every year the government produces statistical data and commentary on demographic and social trends in its official publication, *Social Trends*. Its working definition of the family (which is needed to help categorize the population statistically) is as follows:

> A family is defined as a married or cohabiting couple, with or without their never-married children (who have no children of their own), or a lone parent with such children. People living alone are not considered to form a family.
>
> (*Social Trends*, 1999, p.43)

The cartoons in Figure 2.2 present a vivid and amusing representation of family life today which is almost certainly recognizable to you (in a way which it would not have been to people say 50 years ago). The top one is close to what Judith Stacey (1991) calls 'brave new families', whilst the bottom cartoon captures the diversity of living arrangements now possible. But what 'hard' data do we have? What do statistics tell us about family life today and how it has changed during the past few decades?

FIGURE 2.2 Family life today?

Source: *The Observer*, 25 October 1998

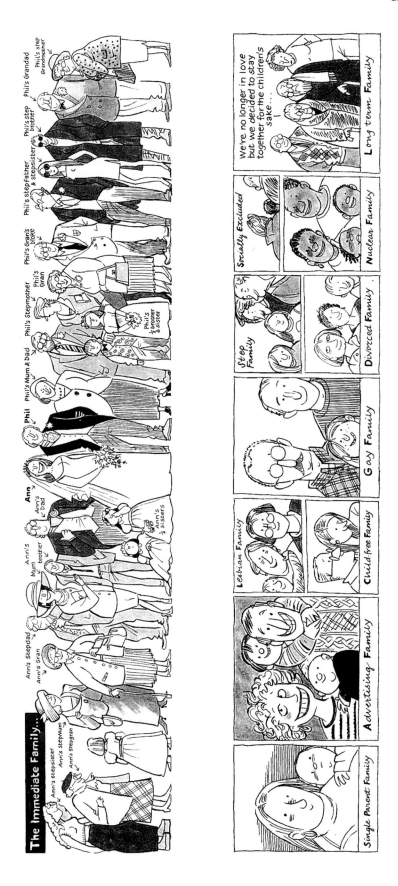

ACTIVITY 2.2

The following tables and graphs are taken from or adapted from *Social Trends* (1999). Between them they allow us to identify some of the changes and continuities in family life since the 1960s (or in some cases the 1970s). They also allow us to begin to address questions of how diverse, and how uncertain, family life is today compared with several decades ago. But first you should look at each table/graph or pair of tables/graphs, and jot down notes to the questions directly beneath.

(Please note where figures have been rounded up to the nearest final digit, there may be an apparent discrepancy between the sum of the constituent parts and the total as shown. Note also that references are sometimes made to either the UK or Great Britain. This is due to the different surveys from which the data are extracted by *Social Trends*.)

TABLE 2.1 Households by type of household and family, Great Britain

	Percentages				
	1961	1971	1981	1991	1998
One person					
Under pensionable age	4	6	8	11	14
Over pensionable age	7	12	14	16	14
Two or more unrelated adults	5	4	5	3	3
Single family households[a]					
Couple					
No children	26	27	26	28	28
1–2 dependent children	30	26	25	20	19
3 or more dependent children	8	9	6	5	4
Non-dependent children only	10	8	8	8	7
Lone parent					
Dependent children	2	3	5	6	7
Non-dependent children only	4	4	4	4	3
Multi-family households	3	1	1	1	1
All households (=100%) (millions)	16.3	18.6	20.2	22.4	23.6

[a] The term 'single family households' is used to distinguish them from households which consist of more than one family. A 'single family household' may mean a couple, or a lone parent.
Source: *Social Trends*, 1999, p.42

TABLE 2.2 Percentage of dependent children living in different family types, Great Britain

	Percentages				
	1972	1981	1986	1991–92	1998
Couple families					
1 child	16	18	18	17	17
2 children	35	41	41	37	37
3 or more children	41	29	28	28	25
Lone mother families					
1 child	2	3	4	5	6
2 children	2	4	5	7	7
3 or more children	2	3	3	6	6
Lone father families					
1 child	–	1	1	–	1
2 or more children	1	1	1	1	1
All dependent children	100	100	100	100	100

Source: *Social Trends*, 1999, p.43

1 Using Tables 2.1 and 2.2 can you identify the most significant changes in types of family between 1972 and 1998? You should also try to identify some *continuities*.

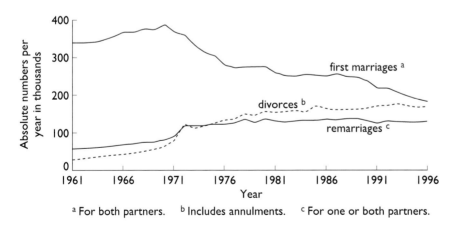

a For both partners.   b Includes annulments.   c For one or both partners.

FIGURE 2.3 Marriages and divorces, UK
Source: *Social Trends*, 1999, p.41

2 What are the overall trends between 1961 and 1996 shown in Figure 2.3 in relation to:
   • first marriages,
   • divorce, and
   • remarriages?
   Can you offer an explanation for the sharp increase in the divorce rate in the early 1970s?

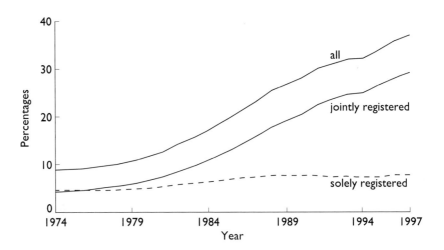

FIGURE 2.4   Births outside marriage as a percentage of all live births, Great Britain
Source: *Social Trends*, 1999, p.50

3   From Figure 2.4, what is the overall trend in births outside marriage between 1974 and 1997? What is the significance of the difference between jointly registered and solely registered births?

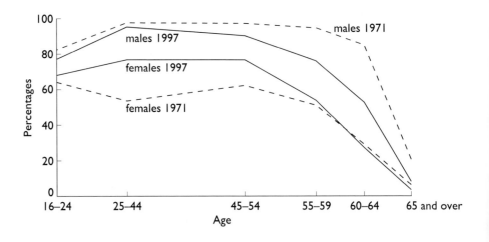

FIGURE 2.5   Economic activity rates by gender and age, UK, 1971 and 1997 (the percentage of the population that is in the labour force)
Source: *Social Trends*, 1999, p.71

TABLE 2.3 Economic activity status of women by marital status and age of youngest dependent child, UK, 1998

	Age of youngest dependent child				No dependent children	All women
	Under 5	5–10	11–15	16–18		
Not married/cohabiting						
Working full time	9	18	30	44	47	39
Working part time	18	31	30	30	20	21
Unemployed	9	9	7	–	5	6
Economically inactive	64	41	33	20	28	33
All (=100%)(millions)	0.7	0.6	0.3	0.1	4.4	6.1
Married/cohabiting						
Working full time	20	26	36	40	49	37
Working part time	37	48	42	41	26	34
Unemployed	4	3	2	–	2	3
Economically inactive	40	23	20	17	23	26
All (=100%)(millions)	2.4	1.6	1.2	0.5	5.2	10.9

Percentages

Source: *Social Trends*, 1999, p.73

4 From Figure 2.5, what was happening to economic activity rates for women in relation to men between 1971 and 1997?

From your reading of Table 2.3, showing data for 1998, how is work outside the home for women related to motherhood?

C O M M E N T

Of course these tables and figures are only a selection of the range of data available relating to changes in family life. You will have the opportunity as the chapter progresses to explore others. But these data do enable us to focus on what seem to be the main changes which may be making family life at the beginning of the century very different from family life even 30 years ago and to consider also those continuities that seem to exist alongside the changes.

As we can see from the first two sets of data (Tables 2.1 and 2.2) the proportion of 'traditional' households comprising a couple with dependent children has been falling since the early 1960s. From making up 38 per cent of all households this figure had fallen to 23 per cent by 1998 (from Table 2.1: percentages arrived at by adding the figures of those with

1–2 dependent children and 3 or more dependent children). Over the same period there has been an increase in the percentage of households consisting of a lone parent with dependent children, with 21 per cent of all children in 1998 living in lone parent families (Table 2.2: lone mother plus lone father families). Almost certainly this will not surprise you. For much of the 1990s lone parents have been a target for political and media attention. But it is also important to recognize the degree of continuity these figures indicate with four out of five of all dependent children, in 1998, still living in a family with two parents (Table 2.2: all types of couple families).

One of the most important explanations for the increase in lone parents is the rising divorce rate. (The most common route into lone parenthood in 1998, in 72 per cent of cases, was the break-up of a partnership compared with 18 per cent due to the birth of a child outside a partnership; *Social Trends*, 1999.) Figure 2.3 shows how the divorce rate has been rising steadily with a particularly marked increase in the early 1970s following legal changes which not only made divorce easier but also started a process of making it more acceptable. But what Figure 2.3 also shows is that many of those divorced marry for a second time. First marriages may be on the decline but marriage continues to be popular. And this has implications for children of divorced parents. According to a recent estimate (Office for National Statistics cited in *Social Trends*, 1999) almost 1 in 4 children born in 1979 in England and Wales was affected by divorce before reaching the age of 16, continuing family life either in lone parent families or in step families.

Even the data on births outside marriage (Figure 2.4) presents us with the same mix of continuity and change. Over a third of all live births in Great Britain in 1997 occurred outside of marriage, more than four times the proportion in 1974. But four out of five of these births were jointly registered by both parents; and according to further data three out of four of these to parents living together at the same address. Only in the case of births to teenage mothers was a significant proportion (29 per cent) solely registered (*Social Trends*, 1999). These figures seem to be showing a turning away from marriage towards cohabitation rather than a turning away from shared parenthood.

Figure 2.5 and Table 2.3 describe a slightly different set of changes, but ones which are bound up inextricably with the others we have been looking at in this section. The economic activity rates for women in 1997 are on average 20 per cent higher than they were in 1971, whilst the rates for men are lower than they were in 1971. This increase in women's activity rates is accounted for by the increased percentage of women with children who are now in paid employment outside the home. As you can see from Table 2.3, women with very young children, whether lone parents or part of a couple, are still more likely to be economically inactive. (And this continues longer for lone mothers.) Furthermore, the majority of women work part time until the youngest child has reached school leaving age.

These figures reflect women's continuing responsibility for child care. But it is a responsibility which now goes alongside a responsibility for contributing significantly to family income: a degree of change which must enter into our discussions in this chapter of the power relations between men and women.

We do need to be careful about exaggerating claims about the differences between family life in the 1950s and today. In the first part of this section we distinguished family life as it was lived in the 1950s from the idealized version so eagerly promoted even today. And we have now just seen how statistical trends describing changes over the past few decades are themselves somewhat contradictory, showing signs of both continuity and change.

However, in spite of this qualification, we do have evidence which suggests that family life in the UK today is different from that of post-war years and that it is its diversity which makes it particularly different. Each of us throughout our lives may now live in a number of different family types and our family relationships may take a variety of forms. This suggests a marked contrast with the 1950s. Just as the 1950s were characterized by the dominance of one particular type of family consisting of husband and wife and children living together in the family home (sometimes labelled the nuclear family), with husband/father as breadwinner and wife/mother caring for home and children, so the beginning of the twenty-first century is characterized by a diversity of family types and a diversity of relationships within families. Lone parent families, step families, families with unmarried parents, families headed by same sex couples exist alongside the traditional nuclear family. And families in which women stay at home to care for young children (or elderly relatives) exist alongside those in which women may be working long hours outside the home in paid employment. With this diversity goes choice. We are probably justified in saying that there did not exist the opportunities in the 1950s that exist today for people to choose how they want to live. Although the two-parent family remains numerically dominant today, this institutional form no longer defines so exclusively what it is to live in a family or what a family is.

We have seen from the above evidence that there may be increasing *diversity* in family life today. But what evidence is there from these facts and trends that family life is becoming increasingly *uncertain*?

Again we do have to be careful about assuming that every family in the 1950s conformed to the idealized model of the nuclear family. Uncertainty, whether resulting from dependence on just one wage earner or from the significant inequalities of power within families, may certainly have been part of family life for many at that time. But there was an expectation of stability which we may not have today. During the 1950s, for example, young men and women on their wedding day expected (though even then it was not certain) that their relationship would continue until one of them died. Children expected to live with their mothers and fathers until they left home. Mothers and fathers knew what was expected of them; they had clearly defined roles. These certainties arguably no longer exist for many of us. In particular, step mothers and fathers, brothers and sisters, from being the stuff of fairy stories, have become for many people a part of the new ordering of family life. The youngest bridesmaid in the cartoon in Figure 2.2, for example, may well have a less certain picture of her relatives than her counterpart of 50 years ago. But this uncertainty does not mean a lack of ordering, just a different kind of ordering, dependent still on the central place the institution of the family has in our understandings of how to live.

SUMMARY

- The main features to the nuclear family of the 'golden age' are predicated on seemingly quite clear-cut lines of power. There is the strict and unequal gendered division between husband as breadwinner/head of household and wife as dependent 'home maker'; an expectation that marriage is for 'life'; and that the normal family was that of a heterosexual, marital couple with dependent children.

- The major demographic and social changes in families and households over the post-war period in the UK are associated with increasingly diverse living arrangements now possible. This increasing diversity of forms of family life is illustrated by the growth of lone parent households, the numbers of dependent children living in different types of family, the growth in divorces and remarriages, and the changing rates of economic activity among men and women.

- There are some continuities as well as discontinuities with the past in terms of family life. It is clear, for example, that most dependent children today still live in a family with two parents.

- There is evidence of significant changes pointing to the possibilities for greater choice, diversity and fluidity in the forms of family living as well as the likelihood of greater uncertainties and new constraints around family life.

3 FAMILY ROWS: THE IDEOLOGICAL DEBATE

Think back to the changes we outlined in Section 2. Do you welcome the increased diversity of family life today? Or do you have concerns? Is the family becoming more fragile? Is it under threat? Or do the changes we have been looking at suggest a strengthening of family life? Are we seeing just the latest in a long history of the reordering of family lives over the twentieth century?

There are a range of possible answers you could have given to these questions. Differences in age, religion, ethnicity and class could all produce different responses and your answers will reflect more than simply your views of family life. They will reflect your wider set of beliefs about the nature of society and how it should be ordered. And this is not just the case for ordinary people, for discussions we might have in the workplace or at home. It applies equally to all those commentators employed to think or write about the family today.

The concept of a *political ideology* is often used in the social sciences to help us explore these wider sets of beliefs. The word *ideology* itself is used in different ways in the social sciences but we are using it here to mean:

- a coherent way of seeing the world (a world view) which is fundamentally shaped by who we are and our place in the social order.

The word 'view' here implies just one angle or standpoint amongst many. However, specifying that they must be *coherent* clearly indicates not just a random collection of opinions but a way of seeing the world in which different elements connect with each other and add up to an overall understanding which holds together. For our purposes here we can take the term *political* simply to mean 'concerning power'. So political ideologies are different clusters of understandings about social structures and institutions, including ideas about power (how it works and usually how it *should* work). The word 'should' also signals that political ideologies are inescapably concerned with social values, and with making judgements about what is better than something else and why, be it the nuclear family, full employment or the welfare state.

Using political ideologies can therefore help us to explore the ongoing debate about what the family is and what the family ought to be (Silva and Smart, 1999). In this section we look at the influence of two contrasting political ideologies, conservatism and feminism, in shaping how we think about the way ordering in families occurs and how we make sense of contemporary changes. Your own responses to the questions at the start of this section may well have reflected increasing public concern about transformations in family life over the past decades and with the increasing diversity identified in Section 2 being viewed as harmful and synonymous with disorder. The assumption here is that the family as an institution should not change. These views, most associated with the political ideology of conservatism, are however open to challenge and contestation from other ways of thinking, not least from the political ideology of feminism. So your responses may also have reflected public arguments about the need to take change, fluidity and diversity in family arrangements seriously as a process of reordering of family life. Change, according to these views, is not necessarily dangerous and undesirable.

Now let's look in some detail at these two opposing ways of explaining why and how family life in post-war UK society is being transformed. We will also explore how the differences between the two political ideologies stem from the differing social values about what is good and bad about particular institutional arrangements of power for ordering our lives.

3.1 Conservative arguments

The political outlook of conservatism springs from a desire to conserve what exists. Conservatism as a clearly distinctive political ideology emerged in part in the late eighteenth century in reaction to the project of the French revolutionaries. However, this cluster of ideas was also the outcome of deep currents in European thought which rejected abstract reasoning as a method for understanding the human world.

Conservative thinkers since have tended to define what the family should be in terms of a heterosexual conjugal unit based on marriage and co-residence. A clear segregation of tasks based on sexual differences is seen as the 'traditional', 'natural' and 'god-given' way of ordering our lives. It is assumed that the man is the 'natural' head of the family. The family's key tasks are the reproduction of the next generation, the protection of dependent children, and the inculcation of proper moral values in children. The family also disciplines men and women in economic and sexual terms: it keeps us in our proper place. Order, hierarchy and stability are seen as the key features of the 'healthy' family and the 'healthy' society. But conservative commentators emphasize the gap between what 'ought' to be and the actualities of how the family works. What 'should be' is juxtaposed with statistics on lone parenting, divorce, high (male) unemployment, delinquency and crime to draw up a picture of the family in decline (Morgan, 1995; Phillips, 1997). Accordingly, the breakdown of the traditional family is seen as one of the main causes of the claimed wider moral decay in society.

For conservatism, the internal stability of the family is important because it is a – some would say the – key unit in securing the stability of the larger society. Societies are viewed as potentially highly unstable. There are deeply conflicting interests between people, partly resulting from profound differences in people's abilities, their wealth, their status, the freedom they enjoy to meet their needs, and so on. As a result, there are risks of envies, rivalries and resentments which are potentially explosive. History has no shortage of examples of internal strife between conflicting interests in the same societies, and the results are usually damaging for all concerned. Left to their own devices, people by 'nature' are undisciplined and self interested, driven more by passions and ambitions than by reason. The exercise of authority and discipline by leaders, by a strong state and by a firm government are the first bulwark against potential conflict and chaos. The second bulwark is the family, in which most human needs can be managed, and future citizens can learn self-control and respect of rules and authority. On the whole, societies like the UK have held together on this basis. The precise function of every social institution, rule and relationship is too complex for anyone to understand (in this sense social scientists are viewed with scepticism) and by far the best strategy is to avoid significant change until the effects on the complex social organism are understood. The allegedly destabilizing effects of change in the family are living witness to this, for conservative thinkers.

One of the changes which concerns conservatives today relates to the role of men in families. According to conservative thinkers, the family is ideally ordered around a dominant male figure, who provides economically for, and holds power over, other members of the family. If we follow this argument, the loss of male power to continue in this position, particularly because of changes in the labour market, the growth of women's employment and the interventions of the welfare state, is fundamentally altering and undermining the secure and hierarchical order which the family creates for the wider society.

ACTIVITY 2.3

During the 1980s and 1990s men's power in families and the extent to which family life can and should be ordered around a dominant male figure has become a focus of interest for academics and the media. This period has seen an upsurge in writing about men and particularly about fathers by academics and journalists, accompanied by the emergence of men's groups concerned to explore their role. The article below is one such example. It appeared in a slightly longer form in *The Observer* in 1997 as an edited extract from the author's forthcoming book, at a time when the Labour government was pursuing policies of moving women 'from welfare to work'. It presents a strongly expressed example of one set of conservative ideas about fatherhood and the ordering of family life.

Read the adapted version of 'Death of the Dad' reproduced over the page then answer the questions which follow.

Death of the Dad

by Melanie Phillips

... there are profound sexually based differences between mothers and fathers. Motherhood is a biological bond fuelled by hormonal and genetic impulses. Fatherhood, on the other hand, is to a large extent a social construct, but founded – crucially – on a biological fact. The traditional family is based on retaining the father's continuing contribution in a social contract between the two, founded on mutual interest.

Of course, there are overlaps. Many fathers are only too delighted to nurture their children; and not all mothers feel much maternal instinct. But in all societies that work, the well-being of mother and child depends on paternal investment in nurturing, supporting and protecting the mother-and-child bond.

If there is no imperative for men to undertake this role it will atrophy, to the inestimable damage of children and mothers.

Societies that work most successfully give men a role as principal family provider and protector, which makes them feel not just valuable but invaluable. Once their wives become mothers, the father's role is to support that mother-and-child unit, even if the mother is also in paid employment. And that is because caring for her child is her principal role, in which she needs considerable practical and emotional – quite apart from financial – support.

Male breadwinning, as the American sociologist David Blankenhorn remarks, is neither arbitrary nor anachronistic. It is important both to cement masculine identity and to civilise male aggression. That is why unemployment has played havoc with young boys' socialisation and shattered their fathers' emotional and physical health.

Employment is an instrumental, goal-driven activity, which permits men to serve their families through competition. It directs male aggression into pro-social purposes. That is why employment is a fundamental means of integrating men into family life. ...

Britain faces a growing crisis among men. The fragmentation of male identity, caused by both unemployment and the progressive and willed destruction of fatherhood, is creating widening spirals of despair, irresponsibility and violence among men and boys. Boys are under-achieving at school. Men's groups are springing up to articulate a sense of mounting grievance. More and more young men are displaying less and less commitment to their young. Thanks to the wonders of reproductive technology, women can now do without a male presence altogether. ...

Female independence can never be bought at the price of the emasculation of men. If mothers wish to work, they will have to devise solutions to child care that do not destroy masculine identity and socialisation. The consequences of destroying the male role of provider and protector are licensed irresponsibility and the promotion of a profound and growing rage, with deeply alarming implications for the social order.

Source: *The Observer*, 2 November 1997; from Phillips, 1997

1 Melanie Phillips has a clear set of ideas about what fatherhood should entail. What are these?

2 She sees fatherhood, defined in this way, as under threat from both male unemployment and increasing female independence. What does she consider to be the *consequences* of these threats?

C O M M E N T

1   We are presented here with a powerful cluster of ideas in which
fatherhood, masculinity and breadwinning are closely linked. For
Melanie Phillips fatherhood is part of a balance of responsibilities – and
so power relations – between mother and father based on mutual
interests whereby the father nurtures, protects and supports the mother/
child bond, providing practical, emotional and financial support for
mother and child. This involves, most importantly, supporting the family
through his paid employment outside the home. But their breadwinning
role not only integrates men into family life. It also enables men to
cement their masculine identity by directing male aggression or need
for competition into a positive purpose, namely that of serving their
families.

2   According to Phillips, threats to fatherhood have consequences for men's
physical and mental health and for their sons' socialization. Destroying the
male role of provider and protector means the 'fragmentation' of male
identity. This results in not only despair and anger and even violence
amongst men but also less commitment to family life and their children
and this has 'deeply alarming implications for the social order', a statement
which of course takes us straight into the concerns of this chapter. Clearly
identified as a conservative world view, it suggests not only fundamental
changes in the way families are ordered but also the destruction of much
that holds the social order together.

It is important to note that this is an extract from the writings of a journalist
not an academic social scientist, with consequent differences in the way the
argument is presented. In particular it is a piece of writing which is explicitly
concerned with what *ought* to be happening and it lets us see very clearly
how social values are part and parcel of political ideologies. We have a
strong sense here of the processes of *interpellation* (that is, of being
recruited into an identity) which can be involved in the formation of men's
identities (**Woodward, 2000**). But note too how this cluster of ideas informs
us about what being a real father entails and the implications of changes to
this role. These social values are illustrative of key aspects of conservative
thinking. The mother and child 'depend' on the 'support' and 'protection' of
the father as 'provider', and fundamental changes in these power
relationships within the family have implications for the stability of society
as a whole. Phillips, in common with all those writing from a conservative
position, does not look forward to new forms of ordering which may
emerge from changes in the roles of fathers but rather sees these changes as
destabilizing and to be avoided since they imply the breakdown of order.

3.2 Feminist arguments

In contrast to the conservative argument about the 'decline' of the traditional family there are responses which see the changes in contemporary family life in a much more positive light. Such responses welcome them because they go some way to redress the inequalities of power relations which are seen as deeply embedded in traditional family structures and in societies in which the 'traditional family' is seen as the norm. This perspective starts from the possibility that the idyll of the traditional nuclear family hides a range of internal divisions, inequalities and asymmetries of power. It suggests that families are, and have always been, not just the source of comfort and support but also the source of oppression for some of their members. Furthermore, this perspective argues that calls for a return to traditional family life need to recognize this.

Again there are many different strands to this kind of response but the most influential ones have developed out of feminism. It was feminist critiques of the family emerging in the 1960s and 1970s, from within the developing women's movement, which drew attention to unequal power relations between men and women within the family. Feminism does, of course, encompass a wide range of positions. It is both an academic approach and a political movement. There is no one feminism and, clearly, it has developed and changed since the early days when feminists first wrote of their experiences of family life. But common to all feminist analysis is a critique of **patriarchy**: that in all spheres of life men have power over women. Although this may not be the case for individual men in relation to individual woman, still society as a whole is characterized by unequal power relations. As you will see below, feminists differ in terms of where they locate the sources of this oppression and in particular what role they see the family as playing in creating and maintaining these inequalities. Even for those whose main focus of interest is the family, their approaches and starting points differ. But common to all is the recognition that the picture of sharing, companionship, and equality presented by many in the 1950s and 1960s actually masked a situation in which men had power over women. From this broad set of ideas and social values, there is the recognition that the increased diversity and difference evident today is to be welcomed as a step towards greater equality for all family members.

Different strands of feminism explain women's inequality in different ways and the role of the family in these explanations also differs. The main concern for *Marxist feminists*, for example, is with the relationship between women's role in families and the working of a capitalist economic system. The reproduction of the workforce (and by this we mean not just the responsibility for the next generation of workers but also the 'servicing' of the needs of the present generation) is carried out by women in the home whilst at the same time women are available to meet changing demands for labour within the economy. Women's work in the home and as a 'reserve army of

Patriarchy
A social system of unequal power relations between men and women, where men exercise power over women.

labour' is, consequently, for Marxist feminists part of capitalist economic relations; it serves the interests of capitalism. For *radical feminists*, the focus is on the way family life enables men to gain and maintain power over women. It is not capitalist economic structures which are the starting point here but the ubiquitous power structures found in every society, whereby men oppress women and which radical feminists consider the most important division in any society. Structures and distributions of power within the family keep women in a subordinate dependent position in the family. This oppression is then played out in a range of other spheres including, for example, the arts, politics, and the world of work. *Liberal feminists* certainly recognize the inequalities that exist in family life and the way these are related to a lack of equal opportunities in many spheres outside the family. However, their starting point is the actions and attitudes of individuals or groups of men and women, which are amenable to change through, for example, legislation or educational programmes rather than the structures of capitalism or patriarchy.

Although each different strand of feminism has a different starting point they have similar interest in understanding the lived experience of family life for women and the obstacles it poses to female emancipation. Whilst there is recognition of the ways in which women's lives have changed during the past 40 years, particularly with their increased employment outside the home, feminists argue that marked inequalities still exist both inside and outside the home. They point to the unequal division of labour in the home whereby even women who are working full time may still take the main responsibility for all the caring and other domestic work. They also recognize that there is a relationship between the priority given to this domestic role, underpinned as it is by assumptions about women's 'natural' interests and abilities, and women's over-representation in lower-paid part-time work outside the home. Indeed, even if 'private' patriarchy is lessening (an argument which not all would agree with) it is only being replaced by what Sylvia Walby (1991) has called 'public patriarchy'. This is a term used to describe the process whereby women have become employees of the welfare state on a huge scale in paid work characterized by both low status and low pay, and in the same kinds of jobs that they have traditionally done at home (e.g. child and elder care).

Equally significantly, feminists have played a key role in developing our understanding that the family is not a *natural* but a *social* unit. According to Gittins (1985), for example, people live in a wide variety of different domestic arrangements and the range is so great that what unifies them is not immediately obvious. Furthermore, 'what orators say about the family is frequently very far removed from how men, women and children actually live out their lives' (Gittins, 1985, p.59).

Consequently, the only way to understand the power of the idea of a 'normal family', Gittins has argued, is as a key ideological tactic, a tool of public policy, an ideal towards which people are supposed to strive. *Deconstructing*

FIGURE 2.6   People live in a wide variety of domestic arrangements

the family as a natural unit and *reconstructing* it as a social one involves exploring and unpacking its internal structures and functions, its wider economic, political and ideological significance. All these need to be untangled to reveal the power relations of men over women and the patterns of individual costs and benefits operating along gender and generational lines (Segal, 1997). And unpacking these will reveal too the ideas about men's and women's 'natural' roles and abilities which underpin these power arrangements.

Although it is clear that feminism views the conventional nuclear family as a source of injustice, it is less easy for feminists to articulate what the family should be. Again there are marked differences between feminists, from those, including *black feminists*, who argue for more recognition of cultural diversity and of differences among women, to others who see increasing separation between men and women as the main way forward, and to others again who support and are encouraged by the new family forms which are now evolving. Whichever the preferred way forward, however, many feminists welcome the way family forms are being transformed both from within and in relation to wider trends. The increasing diversity of living arrangements is celebrated in so far as it enables more equal and liberated relations between women, children and men.

ACTIVITY 2.4

Read the following extract from the feminist writer Lynne Segal, taken from an article in which Segal provides an overview of how a feminist looks at the family. This particular extract, like Phillips' conservative argument discussed above, focuses on the 'new' stress on fatherhood in the 1990s. In reading this you should compare its argument to that of Phillips and try to answer this question:

> How might the debate on fatherhood be used to bolster a conservative moral backlash against the contemporary diversity of households with children?

Lynne Segal: 'A feminist looks at the family'

At a time when men's hold on their traditional familial and paternal authority is becoming less secure than ever before, the new stress on fatherhood can thus serve very old familial rhetoric: the rhetoric which importantly negates feminist insistence upon the actual contemporary diversity of households with children, whether co-habiting single people, lesbian couples, gay men, women on their own, or women living with friends or other relatives. The force of choice or circumstance – perhaps stemming from sexual orientation, perhaps a response to domestic violence, or from a myriad of other possibilities – which may have led people to live outside nuclear families, can thereby once again be pushed aside in favour of unthinking allegiance to the traditional familial ideal. Before embracing the importance of fathers, therefore, we need to pay careful attention to just how easily the abuse of paternal power has been condoned or denied within traditional family life. There are real dangers that the pro-father, pro-family rhetoric so readily merges with the type of conservative moral backlash to feminism and gay politics which we have seen in recent times. It works through manipulating people's sexual fears and paranoia to stigmatize all over again non-familial sex and relationships – an easy thing to do when the harsher economic climate and the tragic reality of AIDS in the 1980s made many feel more vulnerable and search around for scapegoats.

Mindful of these possible dangers, however, I still believe feminists were right to suggest the importance of men's participation in childcare and domestic nurturing as one – although definitely only one – aspect of forging new, less polarized and oppressive meanings for 'masculinity' and 'femininity'. Nevertheless, in a world where men in general still tend to have more financial and social power than women, we need to tread warily, embracing the importance of fatherhood in ways which do not threaten women and undermine recognition of non-traditional household arrangements.

Source: Segal, 1997, p.309

COMMENT ⎯⎯⎯⎯⎯⎯⎯⎯⎯⎯⎯⎯⎯⎯⎯⎯⎯⎯⎯⎯⎯⎯⎯

According to Segal, the stress on fatherhood may lead to a dangerous celebration of traditional family life, with all its abuses of paternal power, and the stigmatization of both non-traditional relationships and household arrangements. Her argument also questions Phillips' claim that there has been a loss of male power, instead pointing to the financial and social power which men in general continue to possess when compared to most women.

From the discussion in this section it is clear that the political ideologies of conservatism and feminism provide very different starting points for an analysis of the ordering of families today. Even though there may be agreement about the increasing diversity of family life, they differ in terms of their analysis of the traditional family, the reasons for the changes, their responses to the changes and whether these changes represent disorder or reordering. It is important to recognize these differences. Here we do have two fundamentally different ways of thinking about society. And it is part of your job as a social scientist to be aware that these different political ideologies exist, to be able to unpack them, in the sense of exploring key ideas and assumptions, and to understand why, at certain times or with certain groups of people or certain societies, one or other may be more powerful. At the same time we have to recognize that we ourselves belong to and write from within, or draw from, these ideologies with their conflicting social values.

Let's think about how the differences between these two ideologies stem from differing social values. Clearly, conservatism sets a very high value on social stability. Keeping society as one cohesive orderly whole, with all its faults, differences and even minor injustices is worth a great deal more than attempts to create greater equality and justice which are likely to fail, and which risk serious social fragmentation and even anarchy when they do. For feminism, it is this kind of unchanging hierarchical stability which has repressed and subordinated women (and children) for centuries. Treating stability as a primary social value simply legitimizes inequality. For feminists, the pursuit of social justice is a key social value, particularly in respect of affording women autonomy, equal power, fair treatment, respect, and equal human rights. The value conflict is clear between these two clusters of ideas.

SUMMARY

- Conservatism and feminism are two opposing ideological perspectives which attempt to explain the changes in the post-war family in the UK.

- The conservative defence of the traditional nuclear family is in terms of the family's key stabilizing functions and its critique of the 'new' diverse family forms.

- Feminist writing argues that the traditional family is a site of patriarchal oppression.

- Explanations of changing family forms are part of wider clusters of ideas, beliefs and social values about power which we have termed political ideologies.

4 THEORIZING POWER AND ORDERING IN FAMILIES

As you read through the preceding section you will have been aware of the extent to which it is simply not possible to talk about family life without introducing questions of power. Both conservatism and feminism have much to say about power and family life although what they say is very different. Similarly, questions of ordering have also been running through the other sections of this chapter, not least in the recognition that the family operates as an ordered and stable fixture in people's lives but is also a source of instability as it changes in both its form and its substance. But now it is time to address these questions of power and ordering more directly using the conceptual tools provided in Chapter 1. In that chapter John Allen introduced the sharply contrasting theoretical claims and conceptualizations of power associated with the approaches of Weber and Foucault. In this section we focus on what the 'pay-off' from this theorizing is in terms of explaining one particular feature of power arrangements in the family, namely the role of the father as 'head of the household'.

Table 1.1 in Chapter 1 (p.39) presented a way of exploring the ideas of Weber and Foucault around three key distinctions (questions, theoretical claims, and evidence). This table was used to see how the different questions which Weber and Foucault ask about power shape their contrasting explanations of the genetically-modified food controversy. In the case of Weber, the focus of Table 1.1 was on his analysis of formal, bureaucratic power. Remember, however, from Box 1.2 in Chapter 1 (p.31) that Weber explicitly contrasted bureaucratic authority with that of patriarchal authority arrangements most often associated with family households.

ACTIVITY 2.5

In Table 2.4 we have reproduced the key questions raised by Weber's and Foucault's theories of power. Now, on the basis of the work we have done in Sections 2 and 3 of this chapter, you should try and answer these questions by filling in the 'theoretical claims' box with regard to the changing power of the 'traditional' father in the ordering of family lives. (We will come back to the crucial third issue of evidence later.) In trying to fill in the 'theoretical claims' boxes for Weber and Foucault remember to go back to both Table 1.1 and Box 1.2 in Chapter 1 to help you attempt this activity.

TABLE 2.4

	Questions	Theoretical claims
Weber	Who holds power?	
	Who controls the rule-making machinery?	
Foucault	How is power exercised?	
	How does power circulate?	

COMMENT

We came up with the following attempt at applying each of these theories to the study of the changing role of the 'traditional' father figure and power arrangements in the family.

TABLE 2.5

	Questions	Theoretical claims
Weber	Who holds power?	Men 'possess' power over women and children in traditional families due to the personal and hierarchical authority of the patriarchal male head of the household. Men have traditionally held coercive capacity over their dependants as well as economic power as 'breadwinner'. Patriarchal power is a top-down, visible affair supported by taken-for-granted, traditional ideas of dutiful conduct, passed from one generation to the next. However, the patriarchal power of the father is also backed by bureaucratic rules in modern societies, not least through laws which determine a range of family matters from parenthood and benefit rights to the control of sexuality.
	Who controls the rule-making machinery?	Ordering of families appears to involve both bureaucratic and personalized modes of power which may at times be in tension with each other (e.g. modern laws vs. traditional wisdom) and at other times be in collusion with each other (e.g. state legislation and institutional practices which legitimate patriarchal dominance and perpetuate female dependency).
Foucault	How is power exercised?	Domination works on the basis of self-restraint, with the power of the traditional father figure often internalized by members of families as 'natural'. People generally bring themselves to order rather than as a result of any visible, top-down coercion.
	How does power circulate?	Power relations in families, even traditional patriarchal families, circulate between actors, and this process is one of negotiation and contestation. As well as closing down possibilities about how families may be ordered, the circulation of power opens up possibilities of new ways of arrangement of family lives (by new 'diverse' households and their reordering of family lives).

The notes we have made in Table 2.5 show some of the ways in which these two contrasting ways of theorizing power may be applied to just one feature of the traditional family. As we saw in Chapter 1 with the example of GM food,

FIGURE 2.7 Understanding power: division of labour in the household

each theoretical approach has a selective focus. Certain relationships are illuminated at the expense of others. Furthermore, both approaches point in different ways to relationships between things which we may not otherwise have noted. Different concepts are brought together and lead to different directions in our understanding of, in this example, the patriarchal power of the traditional father figure.

So far we have not examined what evidence each approach would look to. We may note from Table 1.1 that the Weberian approach looks to 'visible actions' – in other words such evidence as that found internally in the family in the ways in which the division of labour in households is organized and rule-bound. Similarly, evidence of power arrangements within families would be sought in decision-making processes which men and women engage in, ranging from allocation of the family budget to child-rearing practices, etc. However, the Weberian approach would also look at evidence 'outside' the family itself, such as the visible actions of governing bodies, for example the state and its 'expert' agencies. Here the focus would be on evidence of the dominant rule-making process at work 'from above' *on* families.

By way of contrast, Foucault's approach to finding evidence to support theoretical claims about the workings of power in families is less concerned with clearly 'visible actions' of particular agencies. Rather it looks to more elusive evidence of how power is exercised and circulated in families. The uncovering of accepted 'truths' would be a crucial source of evidence in this approach. For example, the notion of distinct and unequal gender roles between men and women as both 'inevitable' and 'obvious' due to nature or religious traditions would be traced back over history. This approach would also seek out evidence of the contestation and negotiation of power arrangements in both the traditional, patriarchal family and the 'new' diverse households at the beginning of the twenty-first century. Different evidence is brought together in the two theories which again leads to different directions in our understanding of the changing power of the traditional father in the ordering of family lives.

It is clear from the above that the two theories offer, on many counts, competing explanations of how power works in families. Furthermore, each has its limitations and silences. Authority, for Weber, derives by virtue of position and/or by expertise, usually supported by clear-cut rules. In turn, obedience is secured from above by those in authority over those in subordinate positions in the hierarchy. This emphasis runs the risk of downplaying the negotiation and contestation of power arrangements from those not in positions of formal authority.

Who holds power is not solely determined by family roles. Women and children can and do overcome the most authoritarian of patriarchs by persuasion, manipulation or even coercion. Unlike Weber, Foucault does not pay much attention to the key Weberian question of 'who has power?' (over someone else), with domination instead theorized as the orchestration of indirect techniques of control. Given the violence of some aspects of traditional, patriarchal male power over women and children in families (Dobash and Dobash, 1979) this silence over the direct coercion of one person over another is a major limitation for social scientific analysis. While recognizing that Foucault's theory helps us make sense of some of the key ways in which male dominance continues to be 'normalized' in many contemporary families – for example, in the often taken-for-granted knowledge that women are better suited to care for children than men – we also need to be aware that such normalization processes are often supported by the expertise of people who hold hierarchically privileged positions of formal power associated with the state.

The ordering of families is thus often formally assigned, rule-bound and institutionalized in governmental agencies. Witness, for example, the power of such office holders as social workers, health visitors and the police in the regulation of 'normal' and 'deviant' families. To take one illustrative instance, health visitors regulate and oversee the parenting skills of all new mothers and fathers in terms of the healthy progress of babies and infants. And, if necessary, legally-backed intervention in family life (including removal of the child from the home) may occur if the latter is judged to be at risk.

From our brief application of Weber's and Foucault's theories to one specific example, namely the power arrangements associated with the traditional, patriarchal father figure, it is clear that the two theories represent contrasting ways of understanding power and ordering in families. Taken together, both theoretical approaches have opened up a more comprehensive understanding of the family beyond the uncritical celebration of certain arrangements of power as 'natural'. This section has also pointed to the inescapable playing out of power relations in the ordering of any social institution.

SUMMARY

- The theories of power associated with Weber and Foucault indicate contrasting ways of both exploring and explaining the workings of power in social institutions when applied to the 'traditional' patriarchal father figure.

- The strengths and limitations of these two theoretical approaches are associated with the contrasting emphases on the visible, 'top-down' nature of power arrangements for Weber and on the elusive nature of the exercise and circulation of power for Foucault.

5 FROM 'THE FAMILY' TO 'FAMILIES': DIVERSITY, UNCERTAINTY AND NEW DIVISIONS

It is clear that a lot of people's lives are ordered by their domestic living arrangements – but in families, in the plural. On the one hand, the family is the most readily identifiable set of living arrangements in our society; on the other hand, the forms it takes are diverse, and becoming more so. The unit called the family seems to be the predominant way in which most people (though by no means all) choose to live under one roof, provide for each other's emotional needs, manage their sexual activity, conceive and bring up children, and so on. But the notion of the family as a standard, relatively predictable, permanent, monogamous, legally endorsed and heterosexual unit is less sustainable today than in the recent past. An increasingly diverse range of families exists, and the differences between them may represent some quite fundamental challenges to the distinctive form of institutional order known as 'the family'.

The fact of diversity entails a questioning of the ideological prominence of the nuclear family and also raises broader questions regarding the 'natural' roles of women and men. As we have seen in our discussion of political ideologies, such questions raise difficult (and for some uncomfortable and 'dangerous') issues.

Families are no longer just the solid, 'unchanging' bedrock for wider structures in society but rather represent, in Silva and Smart's (1999, p.2) words, 'a re-defined, fluid context for intimate relationships'. In other words, families (same sex partnerships, lone parent families, co-habiting couples with children as well as married couple-based units, etc.) are not just constraining structures, they are also the products of people actively constructing their own intimate living arrangements. As David Morgan (1999) among others has argued, what matters to more people now in our society is what families 'do' as against the increasingly less significant issue of what form they take. And to say 'families order our lives' is a very different claim from 40 or 50 years ago. But to recognize this fluidity and diversity in families is not to deny the continuity of divisions and inequalities in the family.

The following three short newspaper extracts together remind us not only of the divisions between men and women both inside and outside the home but also of the way economic inequalities between families continue and may even be increasing today.

'I concertina my interface with the children'

Happy High-Flyer

David Burton['s] job as an investment analyst requires him to start his day at 7.30 am and leave 12 hours later. At the weekend he catches up on research. ...

The financial rewards are clear – at 39, he earns about £500,000 a year. And Burton, who has two school-age children, admits he would find it impossible if his wife worked as well. ...

He is honest about his ambition, eschewing the standard 'My family are all that matter to me' line of the typical corporate workaholic. But he still applies the language of business to the business of parenting. 'If you concertina your interface with your children, it might actually be better for your relationship.'

Source: *The Observer,* 11 October 1998

'It's cracking us up. I'm so very, very tired'

Weary Wage Slave

... Mrs Ball works from 7 am to 5 pm, seven days a week. Aged 32, she earns £2.85 an hour as a senior care worker at a home for mentally ill people in Hull. Her husband, an engineer, works from 8 am to 5 pm, Monday to Saturday.

'It is cracking us up,' she says. 'I am so very, very tired. ...'

'I love the job, but I'm not doing it as well as I should. I just can't concentrate, and I know the patients suffer.'

'... But the truth is that we need the money.'

Names have been changed to protect identities.

Source: *The Observer,* 11 October 1998

Costly childcare 'keeps mothers out of work'

... For Dawn Benjamin, 29, a single mother in north London, getting childcare has proved impossible. She has a two-year-old daughter, Tavannah, and would like to return to a job in banking.

'I am dying to go back to work. I loved working and want to be a good role model for my daughter,' she said. 'I have tried to get her placed in a nursery, as childminding would be too expensive for me but it has been impossible. Some nurseries say she is too young and the others are full. I want to earn a decent wage to give my daughter a better life. ...'

Source: *The Independent,* 29 May 1999

The first extract presents us with what may look like a picture of traditional family life. There would seem to be a degree of continuity with family life in the 1950s as outlined in Section 2. Here we have a father who is clearly the main, in fact the only, breadwinner for the family. He is dependent on his wife's work in the home to bring home an income which places the family very high on any scale of economic well being.

The second extract presents a different picture, namely a family which is dependent on the wages of two earners. It is very likely Mrs Ball's wages are not as high as those of her husband, reflecting his job as an engineer and hers in 'caring' work. She probably sees herself as supplementing the income he brings home. But she is not working for pin-money (a phrase frequently applied to the earnings of those women who did work outside the home in the 1950s). Her wages are seen as making an essential contribution to the family income. She works because 'we need the money' and in this respect her family is more typical of the ordering of family life today than the family in the first account as more women enter the labour force.

When we move on to one parent families the picture of the ordering of family lives changes again. Lone parents are a diverse group. They include both mothers and fathers, the divorced or widowed, as well as those who have chosen lone parenthood, and their ages may range from teens to fifties. 90 per cent of lone parent families are headed by a woman (*Social Trends*, 1999). The sources and amount of the family income will not all be the same. But even for those parents who have skills which would enable them to earn a good income (and the third extract here refers to a young woman who has experience in banking) the cost of child care means that unless the expected earnings are high, or unless informal child care is available (and this is more usually the case for single fathers), then paid employment may not be an option. In 1996/97 42 per cent of lone parent families in Great Britain were in the bottom 1/5th of the income distribution (*Social Trends*, 1999). They account for a large percentage of those on state benefit who stand at the bottom of our economic ladder.

Here we have representations of three families illustrating aspects of the *economic* diversity that exists today. Differences in family income levels of course are not something which belong only to the beginning of the twenty-first century. They certainly existed in the 1950s with consequences for the life experiences and life chances of the children of that period. However, what is different today is the diversity of ways in which families need to achieve their incomes, and the implications of this diversity. In the 1950s there was very little choice. It was the father who was responsible for the economic support of the family and its well being depended on his ability to bring home a 'family wage'. Today, as opportunities for full-time work in the traditional male areas of employment are replaced by opportunities in those areas once labelled 'women's work' (notably the caring and service industries), families may be supported by one parent or two in any combination of part-time/full-time/short-term work and/or state benefits.

The diversity of sources of income carries with it the possibilities of increasing economic and social polarization between the work-rich households with two adults in full-time, well-paid work, and those families who are dependent on part-time short-term work for whoever can get it (Hutton, 1999). And this diversity carries with it the uncertainty of family income as earners can move, within the space of a few years, between different types of employment or even into dependence on state benefits. That uncertainty is there even for families which seem to be in a strong economic situation. Someone like David, in the first extract, for example, unlike the middle-class breadwinner of the 1950s, will not be looking for a slow steady rise up a secure career ladder with annual increments until his retirement. Early redundancy and the consequent difficult search for a new job are a feature of men's lives today, and someone like David may find it hard to maintain the family income at its present high level. In addition, divorce could mean a significant drop in income level for his wife and children. Even if his wife were able to find child care it seems probable her position in the labour market would be weak because she has been working in the home rather than building up skills and experience in the paid workplace. In our society today it is our position in relation to the market which largely determines our life chances.

However, this uncertainty is much more pronounced lower down the income scale. Family income levels can change easily as parents move between different kinds of short-term work and unemployment. In the past, the security of middle-class families was not matched by some manual workers whose incomes would decline as their physical strength passed its peak. But there was a degree of certainty, of being able to forecast the future, even if that future was a bleak one, which rested on the lack of diversity of employment patterns. That certainty does not exist today. At the lower end of our family income scale are families increasingly at risk of 'social exclusion' (**Mackintosh and Mooney, 2000**) due to these new uncertainties of labour markets – which are explored in Chapter 3.

What seems certain is that the reordering of family lives will take increasingly diverse and divisive forms between households in different economic circumstances. But why have employment patterns changed and what is the state doing to support those families we are now identifying as being economically insecure? Throughout this chapter we have seen an intermeshing of family practices, patterns of paid employment and changing state policies. These crucial links between the institutional sites of family, work and the welfare state are taken up further in Chapters 3 and 4.

SUMMARY

- There is significant diversity today in the way families support themselves economically.

- This diversity can mean increased uncertainty for some families; it also has implications for the ordering of family life.

6 CONCLUSION

The main purpose of this chapter has been to explore how different institutional arrangements of power may be theorized and explained in terms of the changing make up of the family. We have seen that there is both continuity and change in this key social institution over the last 50 years in UK society. Thus we have explored how families are ordered and stable fixtures in people's lives but are also sources of uncertainty and instability for many family members as the institution is transformed in both its form and shape. This reordering (rather than simply 'order') of the family is suggestive of the more provisional nature of this institution in contemporary society when compared to the past.

REFERENCES

Dobash, R. and Dobash, R. (1979) *Violence Against Women*, London, Open Books.

Finch, J. and Summerfield, P. (1991) 'Social reproduction and the emergence of companionate marriage, 1945–59' in Clark, D. (ed.) *Marriage, Domestic Life and Social Change*, London, Routledge.

Gittins, D. (1985) *The Family in Question*, London, Macmillan.

Hutton, W. (1999) 'High risk strategy' in *The Stakeholding Society*, Cambridge, Polity.

Mackintosh, M. and Mooney, G. (2000) 'Identity, inequality and social class' in Woodward, K. (ed.).

Mooney, G., Kelly, B., Goldblatt, D. and Hughes, G. (2000) DD100 *Introductory Chapter. Tales of Fear and Fascination: The Crime Problem in the Contemporary UK*, Milton Keynes, The Open University.

Morgan, D. (1999) 'Risk and family practices: accounting for change and fluidity in family life' in Silva, E. and Smart, C. (eds).

Morgan, P. (1995) *Farewell to the Family*, London, IEA.

Muncie, J., Wetherell, M., Langan, M., Dallos, R. and Cochrane, A. (eds) (1997) *Understanding the Family*, London, Sage/The Open University.

Phillips, M. (1997) *The Sex Change State*, London, Social Market Foundation.

Segal, L. (1997) 'A feminist looks at the family' in Muncie, J. *et al.* (eds).

Silva, E. and Smart, C. (eds) (1999) *The New Family?*, London, Sage.

Social Trends, London, HMSO (annual).

Stacey, J. (1991) *Brave New Families*, New York, Basic Books.

Walby, S. (1991) *Theorising Patriarchy*, Oxford, Blackwell.

Willmott, P. and Young, M. (1960) *Family and Class in a London Suburb*, London, RKP.

Woodward, K. (2000) 'Questions of identity' in Woodward, K. (ed.).

Woodward, K. (ed.) (2000) *Questioning Identity: Gender, Class, Nation*, London, Routledge/The Open University.

Young, M. and Willmott, P. (1957) *Family and Kinship in East London*, London, RKP.

FURTHER READING

On the new, diverse forms of family living in the UK, see Elizabeth Silva and Carol Smart's edited collection of essays, *The New Family?*

If you wish to explore the feminist and conservative perspectives on the debate about the 'decline' of the family, see Patricia Morgan's powerful conservative analysis *Farewell to the Family* and Diana Gittins' highly influential feminist text, *The Family in Question.*

An accessible introduction to the study of the changing family in post-war UK is provide by John Muncie *et al.*'s edited text, *Understanding the Family.*

On power relations in the contemporary family see Graham Allan's sociological text, *Family Life: Domestic Roles and Social Organization* (1985, Oxford, Blackwell).

Work: from certainty to flexibility?

Graham Dawson

Contents

1 INTRODUCTION

TOO BUSY EARNING A LIVING TO LIVE

Flexible work means more work, and it's wrecking our days of rest. Friday has always been a magical word, the sign on the border between work and leisure, between hard work and serious fun.

But the magic is wearing off. For growing numbers of us Friday no longer ends the working week. Saturday and Sunday working is on the increase and for some the seven-day week is the new norm. As the demand for 24-hour services grows, the distinction between days of work and rest is crumbling – and in Britain more so than everywhere else. The British weekend is being quietly abolished.

The weekend is under attack from three different directions: white-collar workers putting in extra hours to get ahead of the pack, lower-paid employees who must do overtime to make ends meet and an increasing number of people doing one job in the week and another at the weekend. There are 1.2 million people in the UK with a second job, almost double the number in 1984. Almost two-thirds of those juggling two jobs are women.

(Adapted from *The Observer*, 11 October 1998)

This chapter explores new ways of ordering the lives of people 'at work'. The certainty of a 'job for life' used to be a reasonable expectation for many people, or at least for many men, looking for paid employment outside the home. Now, a secure job with regular hours cannot be taken for granted; work has become more diverse or flexible with part-time working and short-term contracts increasing in importance. One of the issues raised by this development, indeed the most important one of all, is whether flexible work is beneficial. Is it a new avenue of opportunity or a new source of uncertainty in the ordering of our lives?

From reading the extract at the beginning of this chapter, what do you regard as the benefits and drawbacks of flexible work?

The main advantage of flexible work for workers seems to be that it enables people to bring home more money; people can work the hours that suit them, they can work overtime or weekends, perhaps in a second job, in order to make ends meet. (No doubt most people would prefer to make enough money to live on without having to work overtime or have a second job.) Also, flexible work can be a route to advancement for people in white-collar jobs. The downside is the loss of leisure time, while the phrase

'juggling two jobs' hints at the stresses which can accompany flexible
working. This is especially true for the two-thirds of people in this situation
who are women, given the lack of gender balance in the division of domestic
labour referred to in Chapter 2. Flexibility also means insecurity for some, in
terms of future employment prospects. In short, flexibility produces both
diversity and uncertainty.

ACTIVITY 3.1

Work was not always like this. Have a look at the painting which is reproduced in
Figure 3.1. What can you deduce from it about work and workers in the 1930s?

FIGURE 3.1   Selling the *Daily Worker* outside the Projectile Engineering Works, 1937

COMMENT

To me the main feature of the picture is its marked lack of diversity. The
workers leaving the engineering works are all male, white and middle-aged,
they are leaving at the same time and, since all but one are on foot, they
must live in the same part of town, within walking distance of each other and
the works. There is a feeling of routine and rigid order about the scene,
underlined by the title of the newspaper for sale: the daily worker enjoyed
the security of 'a job for life'. The Projectile Engineering Works looks to me
like a prison, where comings and goings are tightly controlled and the people
inside are not expected to enjoy themselves or do anything particularly

fulfilling. The faces at the window seem to be dreaming of escape – 'a job for life' takes on another meaning. Not all workers in the 1930s were white middle-aged men employed in engineering works. Not everyone in a job today is working part-time or on a temporary contract. But these contrasting images of work dramatize the increasing flexibility and diversity of work, which this chapter will examine.

In Section 2 of this chapter, today's flexible employment patterns and their role in ordering people's lives are placed in historical context. In the 'golden age' of post-war capitalism, the availability of 'jobs for life' in manufacturing industry for many male workers was accompanied by a widespread assumption that most women would spend most of their time at home looking after the children. Section 3 continues the historical narrative with a survey of the extent to which the UK labour market became more flexible in the 1980s and 1990s, focusing on the growth of flexible working practices, flexible employment contracts and flexible pay. These labour market changes have reordered people's lives in important ways (see **Woodward 2000** and Chapter 2 in this book).

Throughout the chapter, three theories of power will be used to interpret this historical narrative. John Allen introduced two of them in Chapter 1. Weber's account of hierarchical power offers insights into the influence of state bureaucracy while it was expanding during the 'golden age' and while it was 'rolling back' its own frontiers in subsequent years. Foucault's picture of more dispersed modes of power also applies to both periods. People internalize assumptions about the 'obviousness' or 'naturalness' of the rather inflexible division of work between men and women that was typical of the 'golden age', and of the more fluid ordering in today's changing circumstances. The third theory of power is rooted in the Marxist account of class structure under capitalism. Each of these theories will be used to explain how power is exercised through the ways work and employment are organized, and how people's lives become ordered and reordered as a result.

This chapter also reflects this book's focus on political ideologies, by introducing social democracy and liberalism. Social democracy was the dominant ideology in the 1950s and 1960s in UK society. Its main values are introduced in Section 2 as background to Keynesian economic theory, widely credited with maintaining full employment until the crises of the 1970s. Section 4 explores the links between the flexible labour market and the political ideology of liberalism, which shaped the UK labour market reforms of the 1980s. As with social democracy, liberal values provide the context for an account of a theory about how the economy works, in this case Hayek's theory of competitive markets.

2 A 'JOB FOR LIFE': FULL EMPLOYMENT AND SOCIAL DEMOCRACY

The aim of this section is to explore the relationship between the full employment and job security of the 1950s and 1960s and the social democratic political ideology which shaped the economic and social policies of the time. Among these policies was demand management, the creation of John Maynard Keynes, which will be explained in Section 2.2.

2.1 Keynes and social democracy

Many economic historians regard the period from the early 1950s to the mid 1970s as the 'golden age' of capitalism. The UK economy grew on average at a rate of 2.5 per cent a year, with only minor booms and slumps around a steady upward path. Unemployment was low, the economy operating at times close to full employment, in marked contrast to the pre-war years. The world depression that began in 1929 reduced the demand for exports from the manufacturing industries on which much of the UK's wealth had been built. During the years between 1929 and 1932, 310,000 jobs were lost in the coal, cotton, shipbuilding and iron and steel industries and the unemployment rate rose from 10 per cent to 22 per cent. This was accompanied by an increase in social inequality, particularly noticeable on a regional level. The industrial job losses were concentrated in Scotland, Northern Ireland, Wales and northern England. Moreover, prices fell during the depression, making people with a secure source of income better off. The greatest concentration of salaried staff and people living on investment income was in south-east England, where the depression was hardly felt at all. It was against this background that John Maynard Keynes set out to rethink economic policy.

No thinker, not even one as innovatory as Keynes, starts from a clean slate, and Keynes was influenced by ideas and values drawn from the social democratic political tradition. He grew up in a household where it seems to have been taken for granted that an intellectual elite, which would guide public opinion on the path to reform, should govern Britain. This was a departure from the liberal idea that the economy would work most efficiently if the state left people alone to pursue their own interests. Keynes was later to express his political beliefs in the following terms:

Laissez faire
A policy of not
intervening in markets.

The question is whether we can move out of the nineteenth century **laissez faire** state into ... a system where we can act as an organised community for common purposes and to promote social justice, whilst respecting and protecting the individual.

(quoted in Moggridge, 1976, pp.46–7)

The phrases 'common purposes' and 'social justice' reflect the humanitarian motivation that lay behind Keynes's work and place him in the social democratic tradition. Social democracy was emerging as a key political ideology in UK politics at this time. It had developed out of socialism, but was distinct from it, particularly in its rejection of revolutionary change and of socialism's strong commitment to replace capitalist private ownership with

Means of production
Factories, machines and
resources used in
producing goods and
services.

widespread public ownership of the **means of production**. But it shared socialism's mission to create a more equal society in which all could prosper, regardless of their origins. Social democracy was committed to social development and social provision to secure common purposes and social justice, through democratic means. This would entail a major role for the state, in the form of some redistribution of wealth through taxation and a system of welfare, and through an active role in planning the economy to secure levels of growth which could sustain greater equalization. Broadly in keeping with this approach, though still very much influenced by his own liberal classical economics background, Keynes was not interested in economic theory as an intellectual exercise but as a way of making the world a better, more secure place. In Keynes's view, policy makers should have warm hearts as well as clear heads and should devote their abilities to 'grappling with the social suffering around them' (Pigou, 1949, p.174).
In the 1930s, this meant above all the elimination of poverty and mass unemployment. It is typical of social democratic thinkers to argue, as Keynes did, that the problems of capitalism could be remedied to produce a better economic system than any other could be. By 'better' here is meant a balance between allowing individuals, including members of the capitalist class, the freedom to pursue their own interests with minimum (but necessary) interference from the state, and ensuring the basic conditions of a secure and certain social life for everyone. For social democrats the point was therefore to 'soften the edges' of the capitalist system; that is, to intervene in the economy to the limited degree that they believed was necessary to eliminate poverty and unemployment. This could be done while leaving the basic capitalist structure, the private ownership of the means of production, intact.

2.2 Keynesian demand management

The root of the problem of unemployment, in Keynes's view, was uncertainty, or more specifically, the volatility of the capitalist system. Business leaders making decisions about investments in new factories and machinery have to take a view about the profits that the new capital goods, the factories and machines, will earn. This means predicting the future course of demand for the consumer goods that their new capital goods will produce. But the future is unknowable:

> If we speak frankly, we have to admit that our basis of knowledge for estimating the yield ten years hence of a railway, a copper mine, a textile factory ... amounts to little and sometimes to nothing.
>
> (Keynes, 1936/1964, pp.149–50)

In circumstances of such uncertainty, investment decisions depended on the mood of the business community, which could swing rapidly from confidence and optimism to caution and pessimism. When that happened, the effects tended to be cumulative: low order books among firms making capital goods led to job losses, the newly unemployed people cut back their expenditure and the fall in demand spread throughout the economy. The essence of Keynes's argument was that the economy could become trapped in a situation of **aggregate demand deficiency**. Even if workers offered to work for nothing, firms would think it pointless to employ them because there was no prospect of selling the goods they would produce.

Aggregate demand deficiency
Too low a level of demand for goods and services in the national economy to guarantee full employment.

This conclusion contradicted the prevailing orthodoxy among economists, which, reflecting liberal political principles, held that unemployment was the consequence of reluctance by workers to accept wage cuts, or, to 'price themselves back into work' in the labour market.

This is how the market works: unless you want to be left with unsold stock, drop your asking price (**Himmelweit and Simonetti, 2000**). The economists who advised the government in 1929 argued that the same logic applied to labour. Unemployment is just another example of excess supply and wages

FIGURE 3.2 A crowd of men gathered at the London docks in the hope of getting work, 1931

are just another price, the price of labour. In order for the unemployed to get jobs, it was necessary, and sufficient, to reduce the asking price – to offer to work for lower wages. One way of understanding how Keynes succeeded in rethinking economic policy is to focus on wages and try to see them in a different way.

From a market point of view, wages are simply the price of labour. How else do we think of wages? What are wages from the standpoint of the wage earner, and his or her family?

To the wage earner, wages are a source of income, in most cases the only source of family income. Without wages, for many families, there is nothing to spend. The significance of this point for unemployment becomes clear if attention is switched from a market, specifically the labour market, to the national economy as a whole. In a sense Keynes 'invented' the national economy and a new branch of economics to study it – **macroeconomics**, as distinct from **microeconomics**, which analyses the component parts of the whole economy, namely households, firms and markets.

Macroeconomics
The branch of economics that studies unemployment, inflation and growth in the national economy.

Microeconomics
The branch of economics that studies the workings of the markets, firms and households that comprise the national economy.

From the viewpoint of the national economy, wages are the most important source of income for consumption; that is, for purchasing goods and services. It is wages that enable most consumers to turn their wants into effective demand in the market. Firms employ labour to produce goods and provide services. Ultimately, therefore, the demand for labour and hence the level of employment depends on effective demand for goods and services. In order to understand Keynes's macroeconomics, it helps to imagine the national economy as a circuit – that is, a circular flow of income (see Figure 3.3). Households receive wages from firms and return most of them to firms in payment for goods and services. Firms use this sales revenue to pay wages to the workers who made the goods, and so on over and over again. The level of consumer demand depends on the level of wages, while the level of employment and hence wages depends on the level of consumer demand. The constituent parts of the national economy are mutually dependent. Change one of them and it has repercussions on the others.

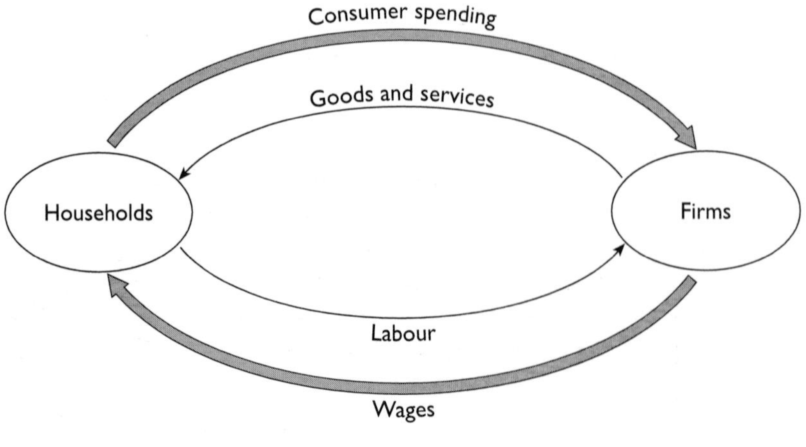

FIGURE 3.3 A simple circular flow model of the national economy

ACTIVITY 3.2

Most economists believed that unemployment was a form of excess supply in a market, and that the cure was to reduce wages. Try to think what, according to Keynes's analysis, would be the effect of a cut in wages for the whole economy.

COMMENT _____

The implication of Keynes's analysis was that wage cuts were worse than unfair or irrelevant – they would actually make the problem of unemployment worse. Lower wages would mean a fall in consumption; that is, a fall in the demand for goods and services and hence in the demand for labour. Wage cuts would lead to further job losses and more unemployment, and would probably also lead to unrest and disorder as occurred in the 1920s and 1930s.

_____

What, then, was the solution to the problem of unemployment, if it was not wage cuts? Keynes believed that it was necessary to consider all of the components of the demand for goods and services, or aggregate demand. As well as households consuming goods and services, firms invest in capital goods such as factories and machines and so they too contribute to aggregate demand. Then there are consumers abroad, who buy the home economy's exports. Finally, but most importantly for finding a cure for unemployment, there is the government, which contributes to aggregate demand through its own expenditure. The government in effect buys the services of doctors and nurses, of police officers and teachers, of members of the armed forces and local authority workers, and provides them to patients, children and citizens free at the point of consumption. The government also buys goods, such as tanks and chalk, and invests in buildings, roads and so on. What matters for employment is the level of aggregate demand. If the government increases its own spending, for example on houses, previously unemployed building workers find jobs and enjoy increased incomes, which they spend on other goods and services. The firms producing them experience an increase in demand and in turn recruit more workers. And so it goes on. The essence of Keynesian **demand management** as a solution to unemployment was therefore to raise government expenditure in order to increase aggregate demand and hence employment. And, of course, the social pay-off was a secure, orderly and peaceful population.

Demand management
Policies that seek to influence the level of aggregate demand.

For social democracy, then, unemployment was not a private misfortune but a collective responsibility. The Keynesian analysis of demand-deficiency unemployment implied that unemployed people were the victims of the uncertainty and volatility of capitalism. In taking measures to increase aggregate demand, the social democratic state embodied the belief that unemployment was the effect, not of a stubborn refusal to accept work at lower wages, but of the malfunctioning of the capitalist economy.

The policies which flowed from Keynes's analysis produced high levels of employment in the nationalized industries and welfare services. Long-term contracts, established jobs, legislation and union power combined so that the expectation of a 'job for life', followed by a company pension, became the norm. If you were in paid work, this was a 'golden age' of security and certainty.

2.3 Perspectives on social democracy

Keynesian demand management was an important attempt to put social democratic ideals into economic practice. How can this project be understood in terms of theories of power? What were its limitations?

2.3.1 Power in the social democratic state

Like all political systems, social democracy and the institutions of the social democratic state created particular kinds of social and economic order. Keynes's theories and the policies they informed had brought about a transformation in the way work and employment were organized, and in the ways in which lives were ordered. And these changes represented important shifts in who held power, how much power and to what ends.

Following Weber's theory, the transformation with which Keynes is associated was a characteristic shift in power at the top which brought about new patterns of entitlement (to work, to welfare) for those who had previously lived in poverty, and new responsibilities (realized through taxation) for those who prospered from enterprise and industry. What brought about this shift were changing ideas, in particular about how the economy worked. The authority of experts (economists and others) to put forward ideas on the basis of impartial rational argument, scientific evidence, and specialist knowledge grew. The appeal to these is clear in Keynes's evidence to the Macmillan Committee on Finance and Industry in 1930. He looked forward to the day when monetary policy 'will be utterly removed from popular controversy and will be regarded as a beneficent technique of scientific control, such as electricity' (Keynes, 1930/1981, p.263). The Keynesian vision of the economy as a gigantic machine, to be supervised and controlled, also fits the Weberian picture. In the 1950s, a Keynesian economist, A.W. Phillips, actually built a hydraulic model to replicate the national economy.

Foucault's approach to theorizing power would bring out some different facets of this Keynesian transformation. Characteristically, it would stress the powers of traditions, assumptions and norms which shape the way people's lives are ordered and which shape how people govern themselves. Whatever the changes Keynesian economic policies brought about, they also left much unchanged. Habitual ways of organizing work continued, not because they were the best or only ways, but because they were entrenched in longstanding patterns of power. Bosses organized work, supervisors monitored the actions of unskilled workers. Married women continued to be largely excluded from full-time paid employment. Discrimination against 'minority' groups kept them subordinate or excluded them. Ideas of gender equality, democracy in the workplace, profit-sharing, or inclusiveness were no more considered after the transformation than before it. Power had changed places in some respects, but not in others.

The Marxist theory of power is especially relevant to the analysis of work. Like Weber, Marx saw social class as a central concept for understanding societies in general, and the distribution of power in particular (see **Mackintosh and Mooney, 2000**). But whereas Weber viewed the economic dominance of owners over workers as one facet of social class divisions, Marx regarded it as the basis of the social structure, from which all other differences and outcomes stemmed. Put at its simplest, the bourgeoisie, who owned capital and the means of production, used their position to exploit the proletariat, or workers, who could live only by selling their labour at whatever market rate happened to prevail. This ownership of capital produces deep social inequalities, serious tensions and frequent conflicts between owners and workers of the kind which occurred throughout the 1920s and 1930s. For capitalism to continue without repeated crises, it was necessary for the state to take on a number of key roles, from 'softening the edges' of an unequal system, to justifying inequalities and maintaining order when instability threatens. Marxist theorists argue that the capitalist class exercises *structural* power over state policy makers (Jessop, 1990; Gill and Law, 1988). According to this theory, the state in capitalist society rules in the long-term political and economic interests of capital. For example, the importance policy makers attach to maintaining business confidence reflects the ability of the capitalist class to shift their investments elsewhere. In this way the state pursues policies which secure the conditions necessary for continued capital accumulation.

The Marxist interpretation of social democracy, and in particular of Keynesian demand management, is that it was a necessary adjustment to the requirements of capital accumulation under mass production. Ford had pioneered the use of assembly-line operation in the mass production of cars, and similar methods quickly spread to the production of consumer goods. Mass production needed a mass market; the vast numbers of goods that could now be produced had to find buyers if profits were to be made and capital accumulated. Mass consumption could take place only if wages were high enough and full employment ensured that everyone who wanted to work was earning such wages. According to Marxist theory, Keynesian demand management did more than alleviate poverty and eliminate mass unemployment. It served the interests of the capitalist class by establishing the conditions for continuing capital accumulation.

2.3.2 'I'm alright, Jack': a postscript to the 'golden age'

Whichever of these theoretical interpretations we follow, what is clear is that, in some respects at least, the balance of power in employment shifted significantly in the 1950s and 1960s. Strikes, demonstrations and workers' militancy before the Second World War underpinned the case for a change to secure jobs and better pay. They also produced a strong sense of solidarity in the working class, particularly in the trade union movement. This often gave organized labour the edge in struggles over pay, working hours and conditions. But for many people the 'golden age' of capitalism was tarnished by the uneven distribution of the fruits of economic growth – good jobs, secure jobs, high wages. Among working-class people, the winners in the struggle to secure a bigger slice of prosperity tended to be white men belonging to trade unions. But there was a downside to class solidarity based upon trade union power, expressed memorably in the saying 'I'm alright, Jack'.

The 'golden age' of capitalism transformed the lives of the men such as those in the picture *Selling the* Daily Worker *Outside the Projectile Engineering Works* (Section 1, Figure 3.1) and in the photograph of men waiting for work (Section 2.2, Figure 3.2). By and large, those without jobs found work and those in work benefited from better pay and working conditions. In Weberian terms, deliberate processes of reordering had increased the power of the most organized section of the working class. But away from trade unionized manufacturing industry, many people were not 'alright'. The social democratic capitalism of the 'golden age' had difficulty in meeting the challenges posed by the growing diversity of the workforce and society in general. Some groups of people were excluded from full participation in economic success. Consistent with a Foucauldian analysis, old habits, assumptions and prejudices were untouched by the transformation which benefited white working-class men. Women participated increasingly in the labour market but mainly in insecure and low-paid jobs (Hatt, 2000). By the mid 1960s, West Indian migration to the UK was almost complete and that from the Indian subcontinent well under way. But migrants were overwhelmingly confined to manual work, frequently in jobs below those for which they were qualified, and were considered by some employers only when no white workers were available (Modood *et al.*, 1998). Furthermore, some commentators identified the emergence of 'the affluent worker' (Goldthorpe *et al.*, 1969) and allegedly growing disparities between the latter and other sections of the population.

At the same time, some politicians, business leaders and economists, who warned continually that the 'golden age' was really a fool's paradise, placed a different question mark over the apparent economic success. Full employment and secure jobs were seen to have created a range of problems. Unions forced wages to levels which made British goods uncompetitive on world markets. Cheap imported goods from countries in which employers could pay their workers less flooded in and well-paid consumers bought them in preference to dearer goods made in the UK. At the same time, it was argued that working practices negotiated by powerful unions resulted in poor efficiency and lower output per worker than foreign competitors achieved. Nationalized industries, in particular, were seen to be inefficient, with management as well as workers lacking the spur of competition from other firms. Price and wage inflation began to spiral, and the trend of steady economic growth began to slow. The costs of maintaining secure levels of welfare benefits, and of the health and education systems, combined to impose a higher burden of taxation on firms and entrepreneurs than in competitor countries. By 1976 high inflation, economic stagnation, a poor export record and high levels of imports of consumer goods brought an economic crisis which forced the government to borrow from the International Monetary Fund to underwrite its debts. The 'golden age' of growth and full employment was at an end.

SUMMARY

- The 'golden age' of capitalism was associated with social democracy, and in particular with Keynesian demand management.

- Keynes analysed the mass unemployment of the 1930s as the product of the uncertainty and volatility of the capitalist system.

- The social democratic capitalism of the 'golden age' succeeded in ending mass male unemployment and promoting prosperity and security, but the benefits of economic growth were not evenly spread.

3   FLEXIBLE WORK IN THE UK TODAY

The period which followed the economic crises of the 1970s saw the collapse of the social democratic consensus partly based on Keynes's policies. The 1980s and 1990s were a period of transformation in the organization and distribution of work and employment. Secure, standard, longstanding jobs and fixed, standard and allegedly rigid working arrangements began to change rapidly. Old values and old assumptions associated with social democracy in general and Keynesian policies in particular were challenged, and supplanted. By the mid 1990s, a central tenet of government policy was that 'flexible labour markets play a key part in a competitive economy' (HMG, 1994, p.50). The assumption is that in today's economic world change is normal: flexibility is usually defined in terms of 'the ability of markets, and the agents that operate within them, to respond to changing economic conditions' (Beatson, 1995, p.1). Many of the changes in the UK labour market have occurred as firms have responded to increasing competitive pressures. The role of government policy has also been important, for example, in aiming to promote flexibility by restricting the power of trade unions. But it has not been a one-way street. In some cases it has been workers who have appeared to want flexible employment, perhaps because they, notably women, had no choice but to combine paid employment with caring labour in the home.

3.1 Flexible working practices inside the firm

Competitive advantage
A cost or quality advantage that enables a firm to be more successful than rival firms.

In the search for **competitive advantage** employers have come to realize the crucial importance of increasing the supply of effort from their workers. Labour's input into the process of production depends in part on the relationships between workers and managers. How efficiently people work depends upon the ways in which managers organize the productive process. The rewards and incentives which managers use to motivate workers, including job security and satisfactory working conditions as well as pay, influence the degree of effort people put into their work. Flexible working is one response to the need to increase competitiveness and it raises questions, not only about efficiency and motivation, but also about power.

ACTIVITY 3.3

Read the following case study of the Longbridge car plant, then answer these questions:

- What do you think is meant by 'flexible working practices'?

- Where is the pressure coming from to make working practices at Longbridge more flexible?

- Apart from lack of flexibility, what seems to be responsible for Longbridge's relatively poor profits performance?

BOX 3.1	**Flexibility cuts both ways: a case study of the Longbridge car plant**

Longbridge is just starting a new stage in its evolution. After months of negotiation, the 14,000 workers at the UK's biggest car plant are expected to agree to flexible working practices already employed at BMW's factories in Germany. The alternative, warned Bernd Pischetsrieder, BMW's chairman and a great nephew of Sir Alec Issigonis who designed the Mini (built at Longbridge), was job cuts, a freeze in investment and the transfer of models to other factories. This would almost certainly condemn the Birmingham car plant to death.

But wait a minute. Surely Britain is supposed to be the country with lots of labour flexibility, compliant unions and free market practices. Germany is alleged to be full of overpaid workers and unions that cling to a 35-hour week. That it is not the way it looks if you compare Longbridge to Regensburg, one of BMW's trio of factories around its headquarters in southern Germany.

A walk down the production line at Regensburg shows how much more flexible Germany's car workers have become. It was Regensburg's workers who pioneered the first of BMW's flexible working time schemes at the end of the 1980s. Out went the standard eight-hour day, five-days-a-week shifts. Instead, workers adopted variable shift patterns, which means they work on average nine hours each day, for four days a week, and are regularly required to work on Saturdays for no extra pay. In this way, Regensburg's expensive machines are turning continually for 108 hours a week (compared with 80 hours a week before the changes). There are no stops for holidays. Moreover, a 'working time account', introduced in 1996, allowed the company to ask workers at Regensburg to work longer hours for no overtime pay during periods when demand is strong and production has to be increased quickly. In return workers can take time off later in the year.

Compare all that with Longbridge. Workers continue to work rigid 37-hour, five-day working weeks, and claim bonuses for late shifts and – a crucial sticking point for the German parent – generous overtime payments.

Improving working practices is however only part of the battle. A chief reason why Regensburg is more productive than Longbridge is that German carmakers work with much better machines. BMW is promising to upgrade the equipment at Longbridge as part of the labour changes. Longbridge produces more cars per worker per year but they are smaller, cheaper and quicker to turn out than the grander machines at Regensburg.

Pre-tax profits at Longbridge fall far behind Regensburg. And that goes for wages too: the base salary of Regensburg workers is £20,000; at Longbridge it is £16,500. But it took the German workers a long time to achieve that higher standard of living. It may take their British colleagues a long time to catch up.

Source: Adapted from Bowley, 1998, p.11

COMMENT

Flexible working practices mean flexible working time schemes (variable shift patterns, working time accounts), making several car models in the same factory and shifting production between plants. The rigidities of the old 'job for life' system which came into disrepute at the end of the so-called 'golden age' are viewed as being at the very heart of the claimed uncompetitiveness of British industry, as they were in the 1970s when the era of Keynesian-inspired policies began to be challenged. In this example, the pressure to make working practices more flexible comes from the higher productivity and profitability achieved in Regensburg. Longbridge's poor profits performance reflects the better machines in use in Regensburg and the more expensive and hence more profitable cars made there.

Let's now think about what motives the workers at Longbridge might have for adopting flexible working practices. And, thinking back to John Allen's discussion of power in Chapter 1, in what ways are the managers at Longbridge (and in Regensburg) exercising power over the workers at Longbridge?

The answer is basically a stick-and-carrot approach: the threat of job cuts, investment freeze and factory closure; and the prospect of higher wages. The BMW managers exercise power over the Longbridge workers through these different ways of motivating them. This seems to be a fairly clear example of power in one of the senses distinguished by Weber: namely bureaucratic or legal-rational authority. Workers are persuaded to obey a set of objective rules which reflect the bureaucratic structure of large organizations such as car manufacturing firms. If we ask where the 'real' power lies, we might want to look beyond a large bureaucracy, in the way Marx suggested. The ultimate test seems to be 'profits per worker', which points towards a relationship of power between workers and shareholders, or between labour and capital. For the Marxist, the main point, and a point of central importance about the labour market, is that this power relationship is structurally unequal. Workers are more dependent on a weekly wage than capitalists are dependent on finding workers. Capitalists have reserves. In particular, capital is mobile, in that shareholders can buy out of one company and into another, and firms can close plants, transferring production to another factory. But workers are much less mobile, geographically and in terms of skills and occupations.

The Foucauldian account of power sees it circulating around society, without being traceable to any identifiable powerful group. At Longbridge, managers are reordering their lives, re-creating themselves by internalizing new assumptions about what is obvious. Seeking flexible working practices might be interpreted as 'waking up' from the 'quiet life' they used to enjoy in days when competitive pressures were less intense. Now they are creating themselves anew as dynamic, innovative executives on a mission to reinvent Longbridge. So it is not only the workers at Regensburg whose lives have been reordered by the loss of Saturdays at home and by having to be flexible about when to take holidays. Workers and managers alike are in the grip of internalized assumptions, ideas that are everywhere in general and nowhere in particular.

3.2 The flexible labour market

In what ways is the UK labour market more flexible than it used to be, say, 40 or 50 years ago? This question can be broken down into three sub-questions, about wages, job security and flexible employment contracts. What has happened to earnings inequality, in terms of gender and skill levels, during this time? Is the 'job for life' extinct? Are part-time and temporary working more common?

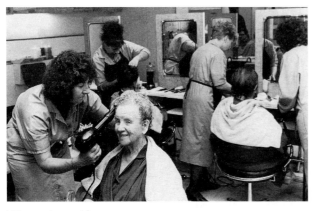

Women's work?

Men's work?

Queen Ella and Prince Charming were living happily ever after...

New times: flexible working?

FIGURE 3.4
Gendered working

3.2.1 Flexible wages

An important aspect of flexible employment is flexibility in wages. To an economist, wages are the price at which labour is bought and sold in the labour market. If you are in work, you sell your labour to an employer. In a competitive market, prices tend to converge to the equilibrium price (see **Himmelweit and Simonetti, 2000**). On the assumption that individual buyers and sellers are self-interested, a competitive market guides them towards the best available deals. In contrast, wages (or prices in labour markets) have traditionally tended to be fixed by agreements reached through **collective bargaining** between employers or employers' organizations and trade unions.

What effects might be predicted to result from increased competitiveness, or flexibility, of the labour market? First, inefficient producers are driven out of business as more efficient producers sell the same goods at a lower price. The implication is that people seeking work will not find it if they are 'inefficient', unless they are prepared to work for lower wages (sell at a lower price). Second, a flexible market provides information to people seeking work. This

Collective bargaining
Wage negotiations between trade unions and employers' organizations rather than individual workers and employers

may make it possible for them to improve their own well-being by selling for more – that is, demanding higher wages, if the information is that their skills are more highly valued than they thought. In other words, just as all labour markets order people into hierarchies in employment by paying them different amounts, more flexible labour markets do so more often and more vigorously. The prevailing ordering is repeatedly being shaken up to produce a new ordering to match changing circumstances. As a result, competition between workers for better paid or more secure 'slots' intensifies. While some people prosper, others slip down the ladder or are excluded altogether. The result, at least potentially, is growing inequalities between workers.

The evidence from the UK labour market suggests that it has become more flexible and that this has had the predicted results, with one important qualification regarding inequalities in pay between men and women to be discussed below. During the 1980s a number of institutional changes to the UK labour market moved it in a more competitive or flexible direction. The incomes policies of the 1960s and 1970s, which had been associated with a compression or equalizing of wage differentials, were abandoned. There was a gradual loss of the power of wages councils to enforce minimum wages in industries such as catering and retail. And the proportion of the workforce belonging to a trade union fell during the 1980s from 58 to 42 per cent.

Wage dispersion
The degree to which actual wage rates deviate from average wage rates.

Union activity has been associated with a reduction in **wage dispersion**, supporting the view that the decline in union presence has contributed to the increasing wage dispersion of the past 15 years (Goodman *et al.*, 1997, p.169). Until recently it was generally taken for granted by some commentators, particularly Marxists, that the only really important source of inequalities in income and wealth was the relation between labour and capital, between workers who 'owned' only their own labour and capitalists who owned the means of production. Research now focuses as much on income inequality between individuals, because in recent years the major changes in the distribution of income have occurred between workers. The 1980s saw an increase in the demand for skilled workers and a consequent rise in their wages relative to those of less skilled and unskilled workers. Since skilled workers were already earning higher wages than unskilled workers, the gap between wages for the two groups widened: that is to say, wage dispersion increased.

The qualification to this story of increasing inequality concerns the male–female wage gap, in part because the institutional changes have been very different. The Equal Pay Act of 1970 and its 1983 Amendment have contributed to some reduction in gender pay inequality. However, women's earnings still lag behind men's. Women's hourly wages only reached 70 per cent of men's by 1993, from 55 per cent in the late 1960s (Hatt, 2000). In 1996 the average weekly full-time earnings of a man were more than £100 higher than those of a woman (Hatt, 2000). This is due in part to the under-representation of women in high-paying occupations, such as doctors and bank managers, compared with the numbers of female nurses and schoolteachers. In 1994, women with A level qualifications were earning only 75 per cent of the weekly earnings of men with equivalent educational qualifications (Hatt, 2000). The explanation of this gender discrimination includes the influence of tradition, trade unions putting a wage premium on skills associated with male employment, and the assumption by employers that domestic work has no skills that are transferable to paid employment. According to Wright and Ermisch (1991), without such forms of economic discrimination women's pay in 1980 would have been 20 per cent higher. Gendered pay differentials remain a source of inequality.

3.2.2 Job security

A relatively high level of job insecurity seems likely to be a feature of a flexible labour market. The evidence on trends in job security supports this prediction and here again a lessening of gender inequality is apparent (Gallie *et al.*, 1998). Let us begin by considering job security for men who were in their twenties during the 1950s. They enjoyed a high level of employment stability but it has declined for men of the same age group ever since then. By the time we reach men who were in their twenties during the 1980s, job insecurity is much greater. At any given time, it is men in their twenties who experience the highest level of job insecurity.

The converse is true of women: employment stability increases as we approach the 1990s. The steepest rise is for women in their twenties and thirties, with women in their twenties experiencing an uninterrupted rise in employment stability from 1966–79. But women in their forties benefit from a higher level of job security anyway. As Gallie *et al.* (1998) put it, the 'overall impression is that women are enjoying more employment stability, while men are finding it more difficult to maintain stable employment patterns' (p.122). However, it is important to record the actual levels of measured employment stability for women in comparison with those reported for men. It is only during the years 1981 to 1992 that the highest level of job security for women (women in their forties) became equal to the highest level of job security for men in their twenties, who have experienced the steepest decline in male employment stability. The position of women is improving relative to that of men, but from a very low base.

3.2.3 Flexible employment contracts

The evidence on flexible employment contracts is complicated by the variety of forms of flexible working: part-time work, temporary work, fixed-term contract work, seasonal work, shift work, flexitime, overtime, zero hours contract employment, term-time only working (as in supply teaching), Sunday working and self-employment without employees. These are all ways of enabling firms to be more competitive in a rapidly changing economy. Firms can adjust the number of workers they have to pay, and the number of hours worked, according to changes in the demand for the goods or services that they produce.

ACTIVITY 3.4

Tables 3.1 and 3.2 have been compiled from Dex and McCulloch (1995). The figures are derived from the Labour Force Survey (LFS), an annual survey of a representative sample of the population of working age which is carried out by government statisticians. Now have a look at Tables 3.1 and 3.2 and answer the following questions:

- What conclusions can you draw about changes in part-time and temporary work?
- To what extent was 'all flexible' employment more common in 1994 than it was in 1975?

TABLE 3.1   Changes in men's employment, 1975–94

	Percentages in each category			
	1975	1981	1986	1994
(A) Full-time permanent	n.a.	n.a.	79.3	73.1
(B) Part-time	2.4	1.7	3.5	6.1
(C) Temporary (full-time)	n.a.	n.a.	2.6	3.9
(D) Self-employed without employees	5.8	4.7	9.0	12.4
Flexible (B + D)	8.2	6.4	12.5	18.5
All flexible (B + C + D)	n.a.	n.a.	15.1	22.4

See notes below Table 3.2.

TABLE 3.2   Changes in women's employment, 1975–94

	Percentages in each category			
	1975	1981	1986	1994
(A) Full-time permanent	n.a.	n.a.	47.7	46.6
(B) Part-time	39.0	40.6	43.8	43.2
(C) Temporary (full-time)	n.a.	n.a.	1.9	2.8
(D) Self-employed (without employees)	2.2	1.5	4.5	5.0
Flexible (B + D)	41.2	42.1	48.3	48.2
All flexible (B + C + D)	n.a.	n.a.	50.2	51.0

Notes: From the 1950s through to the 1970s the government statisticians who compiled the LFS did not feel that it was necessary to distinguish carefully between permanent and temporary full-time employment. Hence the unavailability of the distinction between full-time permanent and full-time temporary employment in the LFS until 1986.

The percentages do not sum to 100, because they omit the figures for people who are self-employed with employees or who are on government schemes.

Source: Based on Dex and McCulloch, 1995, Tables 4.1 and 4.2; data derived from Labour Force Surveys (1975–94)

C O M M E N T

The biggest changes to part-time and temporary work have affected male employment. The data in Table 3.1 show that a higher percentage of the male population of working age was engaged in two of the main forms of flexible employment in 1994 than in 1975. Part-time working was up from 2.4 per cent in 1975 to 6.1 per cent in 1994, while self-employment without employees rose from 5.8 per cent to 12.4 per cent. Temporary full-time working also increased between 1986 and 1994 but the limitations of the data do not permit us to see a long-run trend back into the 1970s. Summing up, flexible employment among men, narrowly defined as part-time employment and self-employment without employees, increased over the whole period. The proportion of men in flexible jobs, or 'all flexible' employment including temporary full-time employment, rose in the second half of the period, from 15 per cent in 1986 to 22 per cent in 1994.

It is worth noting that what might seem at first sight a rather small contraction of permanent full-time male employment – down from 79 per cent in 1986 to 73 per cent in 1994 – hides an unequally distributed impact. Think back to the discussions of the changing order of gender roles and work in Chapter 2 of this book. It is unskilled and low-skilled jobs, manufacturing and extractive industries and hence parts of Northern Ireland, Wales, Scotland, northern England and the Midlands that have been worst affected.

Perhaps the most important feature of Table 3.2 concerns the starting levels of the figures in 1975 for part-time work and hence for flexible employment, and those for permanent and temporary full-time employment in 1986. Flexible jobs occupied a much bigger place in women's employment in 1975 than in men's, and full-time employment (whether permanent or temporary, which we cannot tell) a correspondingly smaller place. In terms of changes over the period, the main difference is that there were only minor changes in women's flexible (and 'all flexible') employment between 1986 and 1994. It seems that, while women are still much more likely than men to be in flexible employment, the gap has become less wide in the 1980s and 1990s.

To sum up, the UK labour market has become more flexible, as measured in terms of wage dispersion, job insecurity and flexible employment contracts. It has also become feminized in the sense that, measured in the same terms, there has been a reduction in gender inequality.

How can these labour market developments be interpreted in the light of Weberian, Foucauldian and Marxist theories of power?

Perhaps the most relevant part of the Weberian account of power is its recognition of what Chapter 1 referred to as the 'plethora of rules and regulations' that public authorities lay down to guide and limit our actions. One reason for greater labour market flexibility in the UK is that legislation to maintain wage levels and job security has been replaced by legislation which has curtailed the powers of the trade unions and the scope of collective bargaining. From a Weberian perspective the flexible labour market can therefore be seen as the creation of the state, exercising authority from above. This is also true of the slightly reduced but maintained gender inequality that has been a feature of the UK labour market. For the Weberian, the effect of equal opportunities legislation is what matters here.

The Foucauldian conceptualization of power as an anonymous force, provoking free agents to act in ways that make it difficult for them to do otherwise, has a certain resonance with the role of competitive markets in spreading information. People tracking labour market fluctuations, assessing how long their skills will remain current and re-skilling as necessary, seem to fit this picture of being agents operating in the light of information that constrains their range of choices.

To Marxists, flexibility is primarily about weakening the power of the workforce, and increasing the profits of capitalists by reducing their dependence on workers and promoting competition between workers. Technological change is also believed to favour skilled workers. Those with the skills to use new technologies become more productive in the jobs they were already doing. When new technology displaces unskilled workers, it also creates new jobs for skilled workers to operate it. Globalization has exposed unskilled labour in the UK to competition from much cheaper unskilled workers in developing countries. And capitalist corporations are thus able to pick and choose where to most profitably locate their operations.

SUMMARY

- Employers exercise power over workers by motivating them to supply greater effort than they would otherwise choose to provide.

- Flexible working practices and flexible pay involve a change in the balance of power in the workplace from workers to managers and ultimately to shareholders.

- In recent years the UK labour market has become more flexible, in terms of flexible working practices, job security and wage flexibility.

- There has also been a 'feminization' of the labour market, with some women's employment opportunities, job security and earnings improving, while those of some men have deteriorated, although women remain subordinate in the labour market.

- All these forms of flexibility represent shifts in power between people, and these shifts are understood very differently by the theories of Weber, Foucault and Marx.

4 LIBERALISM AND THE CASE FOR THE FLEXIBLE LABOUR MARKET

● ●

By the mid 1990s, the organization of work, conditions of employment, and the balance of power between employers and employees in the UK had shifted dramatically from the 'golden age' of the 1950s and 1960s. In part, this was attributable to the economic crises of the 1970s. The old policies could not continue if the state was to avoid recurrent crises. But the determination to find another way, and the particular form (among many alternatives) it was to take was not decided simply on the technical advice of expert economists. Just as the transformation which brought the UK from the recession of the 1930s to the relative prosperity of the 1950s and 1960s was closely associated with one political ideology, so the transformation of the 1980s and 1990s is particularly attributed to another. In this later transformation, some of the fundamental values, ideas and beliefs of liberalism led politicians and economists to make the case for the flexible labour market.

Liberalism starts with individuals and thinks of them as holding their destinies in their own hands. If people seriously want to be rich (or just better off), they can work hard, learn skills, earn more money and perhaps buy shares. A more flexible labour market will help them, by removing obstacles to individual advancement; for example, curbing the role of trade unions to negotiate national wage settlements means regional variations in pay, persuading (some) people to move about, wherever excess demand for skills means higher wages. Attitudes like this inform the liberal case for free markets in general and the flexible labour market in particular.

4.1 Liberal neutrality

For many liberals, the scope for individual freedom of choice is maximized in a society dominated by relationships which do not last very long and where the entitlements and duties of both sides are spelled out in contracts. Many markets are like this. The liberal case for markets is based partly on their efficiency in allocating resources. A flexible labour market is believed to be efficient in allocating skills and expertise where they will be of greatest use to employers and in some cases to society as a whole. Liberals also believe that markets have a beneficial effect on the stability of society. The market is a form of **spontaneous social order**: what holds society together are the reciprocal benefits of contracts between self-interested individuals. But this depends on those individuals being willing, and able, to internalize the norms

Spontaneous social order
The establishment of order throughout society without government intervention.

of the flexible labour market, with limited individual scope for long-term commitments and emphasis on juggling several part-time contracts at once rather than the secure ordering of the past.

Liberalism is one of the most important political ideologies in the contemporary world. It is not a club, membership of which requires allegiance to a carefully defined list of beliefs. It is more like a very extended family of political theorists who share common concerns and values and a vision of an ideal community, but within which there is room for differences of emphasis and disagreements over specific policy measures. The core liberal concern is the protection of the individual from coercion by the social groups to which she or he belongs. The characteristic liberal 'project' has been the demarcation of a private area within which individuals can do as they please, free from interference by the rest of society, represented by the state. Individuals should have rights to certain freedoms: to freedom of speech, of religious worship and belief, of movement and of sexuality; and the right to hold and accumulate private property.

This liberal concern to promote the rights of individuals can be understood in terms of the acknowledgement of the diversity of people's opinions, religious beliefs, ethnic and cultural identities, sexualities, economic position, talents and capacities. On many occasions in history, diversity among people has been a source of conflict. How should the state react to religious, ethnic, class or other conflicts among its people? Should it decide in favour of one side and promote its interests, perhaps imposing it as the 'official' state ideology? Or should it embody in its constitution and laws a plea for the toleration of diverse beliefs, identities and social practices? The latter is the liberal answer and, for many people, it is the toleration and later the welcoming of diversity that represents liberalism's greatest contribution to political thought and human well-being. The implication of this answer is that the state in a liberal society has an essential but limited role.

Liberals tend to assume that conflict, or at least a lack of consensus over goals, is typical of modern societies. The response of the liberal state is to set up procedures that will allow different groups of people to live together, respecting one another's right to pursue their own way of life. So these procedures or arrangements must be neutral or impartial with respect to different social groups and ways of life. Laws upholding the right to various freedoms listed above are among the most important of such arrangements:

> It is not the function of the state to impose the pursuit of any particular set of ends upon its citizens. Rather the state should leave its citizens to set their own goals, to shape their own lives, and should confine itself to establishing arrangements which allow each citizen to pursue his own goals as he sees fit – consistent with every other citizen's right to being able to do the same.

> (Jones, 1989, p.9)

This quotation puts the idea of liberal neutrality in a nutshell. There is the assumption of a lack of consensus; the possibility that, left to themselves, all the citizens will converge on a common set of ends is not considered. There is the defence of individual rights ('to set their own goals, to shape their own lives'), sometimes expressed as a commitment to the value of individual autonomy. And the short quotation begins and ends with the liberal belief in the impartiality of the state. It should not seek 'to impose the pursuit of any particular set of ends', while the right of one individual to pursue their own ends should be consistent with the same right for everyone else.

When these ideals, which inform the constitutional arrangements of many states, are absent, authoritarian governments can do great harm. However, the application of the liberal principles of individual rights and freedoms and the neutral state to the economic sphere raises a new set of difficult and controversial issues. Apart from its key concern with the formation of a modern constitutional state, the early development of liberal political thought was also associated with the rise of capitalism and came to be seen by contributors to other traditions of political thought as the 'official' ideology of capitalist institutions, in particular the free or competitive market. Some liberal theorists saw an affinity between the neutral state and the free market, which came to be advocated as an impartial way of settling competing claims on economic resources. On the face of it, the three features of liberal neutrality in the political realm apply equally well to the economic world. There is the diversity of consumer wants, leaving no obvious way of reaching agreement on what to produce. There is individual autonomy, as producers and consumers make their own decisions about buying and selling. And it seems that there is neutrality, too. For in the absence of a central authority ordering people to produce these goods rather than those, the goods that are produced will not favour any one consumer over any other. Does the free market for labour enable every citizen to pursue their own goals only up to a degree that is consistent with every other citizen's right to do the same? This is the claim that economic liberals have made.

4.2 Liberalism and the restructuring of the UK labour market

Liberalism has played an important role in contemporary UK politics since Margaret Thatcher won the 1979 General Election. The economist Friedrich Hayek was one of the most influential advocates of liberal or 'free market' ideas. In somewhat the same way as Keynes, from within the social democratic tradition of political thought, produced a theory about how the economy works, Hayek constructed, from within the liberal tradition, a theory about how markets work.

For Hayek, markets are a means of communicating information; it is only through the prices that people encounter in the labour market that they can find out what to do, or where to seek work which will be of the greatest benefit to themselves and to the rest of society. Markets are sources of knowledge, which is disseminated through changes in prices. Markets are a way of dealing with the diversity both of people's skills and talents and their wants and needs as consumers – and are also a way of dealing with the lack of knowledge in society about this diversity. Hayek makes much of the uncertainty in our lives; for example, we do not know who is going to use the things we make. But in Hayek's view the market can guide us in the right direction. Elsewhere, he likens the market to a telecommunications system, sending out signals about changes in what people want to buy.

Hayek applied his liberal principles to the labour market in a pamphlet published by the Institute of Economic Affairs (Hayek, 1984). In it he argues that trade unions are the cause of both unemployment and inflation. Trade union policies interfere with the operation of the price mechanism in the labour market, preventing wage differentials from guiding workers to where their skills are most urgently needed. People remain unemployed when, if left to themselves, they could 'price themselves back into work'. Hayek also saw trade union 'pushfulness' over 'excessive' wage rises as the cause of inflation. The extracts from Hayek's pamphlet in Activity 3.5 reveal some of the liberal values and principles that informed his diagnosis of the UK labour market in the 1980s.

ACTIVITY 3.5

You should now read extracts A, B and C from Hayek (1984) and then try to answer the three questions which follow.

A

Each individual can rarely know the conditions which make it desirable, for him as well as others, to do one thing rather than another ... It is only through the prices he finds in the market that he can learn what do ...

All the conscientious devotion in the world will earn nothing for the workers making Triumph motor-cycles if the motorcyclists in Britain and overseas want other cycles.

(Hayek, 1984, p.28)

B

The pursuit of gain is held in bad repute, because it does not have as its aim the visible benefit of others and may be guided by purely selfish motives. Yet the source of the market order is that it uses the immediate concerns of individuals to make them serve needs that are more important than they can know.

(Hayek, 1984, p.33)

C

From a human viewpoint the perhaps most profound advantage of the market over alternative methods of directing the use of resources is that it virtually eliminates the use of force and the coercion of men by other men.

In the real world, [only the market] can know where people are required ... And it is precisely because the decision is not the opinion of an identifiable person ... but results from impersonal signals in a process that no individual or group can control, that makes it tolerable. It would be unbearable if it were to be the decision of some authority which assigned everyone to his job and determined his reward.

(Hayek, 1984, pp.40–1, 54)

1   Why do you think Hayek believes that it is important for the market to be 'free' or flexible? In what way does this underpin his case for restricting the powers of trade unions?

2   What holds society together? How do you think Hayek would answer this question?

3   Liberals believe in the freedom and autonomy of the individual as their core values. What evidence of these values, if any, is there in the extracts above?

C O M M E N T

1   The Hayekian case for the free market is founded upon its role in reducing uncertainty through the spread of information. For example, a fall in the price of a good is, on many occasions, a signal that the supply of it is in excess of the demand for it. It is time for producers of that good to think about switching to the production of something else, which, it is to be hoped, consumers will be eager to buy. It follows that anything that impedes the rapid flow of information through markets undermines the well-being of everyone in society. And the same is true of anything that obstructs or delays people's rapid revision of their plans in the light of new information. This is where, according to Hayek, trade unions distort the reception of price signals in the market. Trade unions would have been failing in their duties to their members if they had shared Hayek's robust attitude to the fortunes of the British motorcycle industry (which collapsed in the 1970s in the face of Japanese competition). Their pressure for government intervention, in the form of subsidies to improve the price competitiveness of Triumph motorbikes or tariffs on imported motorcycles, was intended to block the kind of response to price signals that Hayek thought essential to the efficient use of resources. Moreover, the whole idea of collective bargaining prevents individuals making their own decisions.

2   Extract B suggests that society is held together – people act in accordance with the laws – because this is the way to improve material living standards. This is a purely instrumental and self-interested explanation of

social cohesion; it is believed to be a means to the end of affluence. But there is more to it than that. Hayek does not believe that people reach a conscious decision that this is the way to go. For Hayek, social cohesion just happens, spontaneously, as a by-product of market exchange. It is not consciously planned. Society is held together by people's reliance on each other as producers of the things they need and as buyers of the things they produce. Hayek sees the competitive market as a solution to the problem of social order: what holds society together is the fact that no-one is self-sufficient. If X wants Y to work for her, X must pay Y the prevailing market price, the 'rate for the job'. And the way for Y to earn a living is to make something or provide a service that someone else wants to buy. It is significant that Hayek uses the phrase 'market order' in suggesting that its importance is that 'it uses the immediate concerns of individuals to make them serve needs that are more important than they can know'. So markets have a key role in ordering lives in two ways. They 'make order' out of the apparently chaotic economic activity of millions of independent, individual agents. And they create a particular (hierarchical) order by assigning a value to people's work.

3   The liberal belief in individual rights and autonomy underlies the argument put forward in extract C. For Hayek and many other liberals, the standard way in which an individual's rights can be denied or their autonomy overridden is through the use of force or coercion by the state. In a competitive market the 'coercion of men by other men' – here, as elsewhere, it seems that Hayek was not a feminist – is 'virtually eliminated'.

An objection that is sometimes made against this view of coercion is that it overlooks the way in which 'market forces' cause, for example, the closure of a factory, with compulsory redundancies. The workers made compulsorily redundant might well feel that they have been forced out of their jobs. Indeed, the phrase 'impersonal market forces' is sometimes used to express indignation or resentment at the working of the market: the market is uncaring, indifferent to the hardship and misery it can cause by precipitating the collapse of a firm and the loss of thousands of jobs.

How might Hayek and other liberals reply to this point? This is where Hayek would invoke liberal neutrality. Hayek regards the impersonality of the market as akin to the impartiality that distinguishes the political arrangements of the liberal state. It is this impartiality that makes decisions about factory closures, wage levels and so on 'tolerable'. Hayek's claim is that, for example, the responsibility for the closure of a factory and the consequent redundancies cannot be laid at the door of an identifiable person or group, for the owners were simply responding to price signals that they were not using resources to produce goods that people wanted to buy. Ultimately, for Hayek, all that has happened is that individuals have exercised their right not to buy what they do not want.

While this account of market processes appeals to many people, there are many others who think of themselves as more than consumers searching for the best bargains. They might also see themselves as members of a community that includes the workers in the factory facing closure. On some occasions they might want to act collectively, as members of that community or as citizens of the state, rather than as individuals looking after their own interests in their efforts to order their lives. The desire to find a balance between respecting the legitimate rights of the individual and, when necessary, subordinating the individual to the common purposes of a community takes us away from liberalism and back to social democracy.

4.3 The flexible labour market: power and ordering

In Section 2.2 it was explained that Keynesian social democracy was challenged from two directions: the increasing diversity of society and the perceived need for greater competitiveness in industry. These two challenges make up the historical context in which the flexible labour market emerged. In reflecting on liberal attitudes to social diversity and industrial competitiveness, it is illuminating to consider Foucauldian, Weberian and Marxist perspectives on power.

Liberals regard the increasing flexibility of the labour market as the direct result of human agency or 'power to' in the form of free choices made by agents in accordance with their preferences. This resembles the Foucauldian view of power in highlighting the diversity of outcomes in the flexible labour market. Even among working people, there are winners as well as losers, greater opportunities and rewards for some (skilled workers) co-existing with greater risks and poorer returns for others. However, Weberian analysis also remains highly relevant here. In the view of many social scientists, the distribution of opportunities and risks continues to reflect persistent structural patterns of inequality. The decline in the hierarchical power once exercised by trade unions could be interpreted as an opening up of opportunities for individual skills and talents, effort and enterprise. But Foucault sees power as doubled-edged and he did not think that it could be pinned down to the exercise of freedom of choice by individuals, as many liberals would have it. In making and remaking themselves 'for sale' on the flexible labour market, they do so by internalizing its norms, by conforming to its demands.

Look back at the Longbridge case study (Section 3.1, Box 3.1) and your responses to Activity 3.3 to remind yourself of the pressures that might have influenced the decision by Longbridge workers to accept flexible working practices.

Among the pressures that influenced the decision are the threat of closure, job cuts or an investment freeze and the unfavourable comparison with Regensburg, in terms of productivity and profitability. These pressures towards making one choice rather than another indicate the one-sidedness of many contracts in the flexible labour market. Many workers compete as individuals for work with a single employer who enjoys the freedom to relocate, perhaps to another country with a more compliant labour force. So the flexible labour market seems to be an incomplete institutional expression of core liberal values. The concentration of 'top down' power that derives from unequal ownership of capital means that the flexible labour market is not yet the level playing field enjoined by the liberal ideal of the greatest possible freedom for any individual compatible with the same degree of freedom for every other individual.

The Marxist theory of power recognizes this. It sees power as depending on access to and possession of economic resources (property, capital, labour power). Power struggles are chiefly about class divisions between owners and workers. Faced with the harsh reality of earning a living in the flexible labour market, workers both internalize its norms and experience the impact of hierarchical, 'top down' power. It is that sense of an unyielding reality that is captured by the Marxist account of social structure, where arrangements of power and much of the ordering of lives are experienced as an external constraint on workers' agency and aspirations from owners of capital.

SUMMARY

- The case made by economic liberals for the flexible labour market is linked to a belief in the values of individual autonomy and liberal neutrality as well as to an account of competitive markets as a form of spontaneous social order.

- In the flexible labour market, according to liberalism, individual workers can exercise power to succeed through self-regulation and self-marketing.

- The decline of trade unions has also increased the scope for the exercise of 'top down' or hierarchical power over workers, ultimately derived from the unequal ownership of economic resources.

5   CONCLUSION

How work and employment are organized, distributed and divided is a fundamental determinant of how people's lives are ordered in every society. In most modern societies, labour markets are the key institutions through which this ordering is undertaken. As the Introduction to this book observed, the term ordering is fitting because the snapshot of a particular 'order' of work and employment which prevails at one time is ephemeral. The order which is created today gives way to different order tomorrow. It is not simply that whoever occupies which slots in employment shifts as individuals are promoted or dismissed, or as firms prosper or decline. The whole pattern of who has power in shaping work and employment becomes transformed from one period to another.

This chapter has been concerned with the ordering and reordering of these kinds of transformations, with the continued processes of negotiation which cause frequent changes and occasional transformations, with the political ideologies which bring major shifts in the ordering of work, and with the kinds of economic theories which underpin those shifts. We have looked at the transformation from the free markets, unemployment and recession of the 1930s to the so-called 'golden age' of secure jobs, full employment and fixed practices of the period shaped by social democratic ideology, and informed by Keynesian macroeconomic theory. And we have looked at a second transformation, from this 'golden age' to one of flexible working practices and employment patterns, shaped by liberals, and by Hayek's theories of the market from the 1970s onwards.

Of equal importance to these transformation stories is the power of theory to explain – in this case to explain how power works, and how who holds power is changing. At its simplest, most people who depended on their capacity to work in order to survive were often vulnerable and sometimes powerless before the 'golden age'. The social democratic transformation gave new powers to the workforce, in terms of rights for most, security for many, and collective bargaining power over employers for some. The liberal transformation to flexible employment shifted this balance, reducing security for some employees, empowering many managers, and changing patterns of access for some – especially women – who had been excluded from the labour market. But as the theoretical approaches offered by Weber, by Foucault and by Marxism show, very different understandings are available of how power moves round. The Weberian approach sees changing policies, changing legislation and a series of explicit and rational negotiations deciding how power should be exercised in ordering working lives. The Foucauldian approach sees shifting norms, new orthodoxies and a complex interplay of powerful actors, long traditions and enduring habits shaping change. Marxism sees the continuing efforts of an exploitative elite of powerful and wealthy owners to adapt to changing conditions, in order to maintain their structural advantage and extract profit from workers.

Whichever of these theories is more persuasive, what is beyond doubt is that the change to flexible working has been part of a changing pattern of power and ordering. Flexible working practices may be viewed as having been put in place at the instigation of management, responding to competitive pressures and the need to reward shareholders well enough to persuade them to keep their money in the company. Some forms of flexible employment, such as part-time working and temporary jobs, have enabled some groups of workers to improve their relative positions, even when they still remain in subordinate positions in the labour market overall. Putting it another way, the flexible labour market seems to have shifted the balance of power away from workers as a whole towards managers and shareholders, while further disadvantaging increasingly insecure, unskilled workers. Are these processes of reordering of our working lives likely to continue? The political climate in the UK and in Europe suggests that they will. In June 1999, a report drawn up by the UK and German governments, *The Way Forward for Europe's Social Democrats*, called on centre-left political parties throughout Europe to adopt flexible labour markets and to accept that having the same 'job for life' is a thing of the past.

REFERENCES

Beatson, N. (1995) *Labour Market Flexibility*, Research Series No. 42, London, Department of Employment.

Bowley, G. (1998) 'Flexibility cuts both ways', *Financial Times, Weekend*, 28/29 November.

Dex, S. and McCulloch, A. (1995) *Flexible Employment in Britain: A Statistical Analysis*, Manchester, Equal Opportunities Commission.

Gallie, D., White, M., Cheng, Y. and Tomlinson, M. (1998) *Restructuring the Employment Relationship*, Oxford, Clarendon Press.

Gill, S. and Law, D. (1988) *The Global Economy: Perspectives, Problems and Policies*, Brighton, Wheatsheaf.

Goldthorpe, J., Lockwood, D., Bechhofer, F. and Platt, J. (1969) *The Affluent Worker in the Class Structure*, Cambridge, Cambridge University Press.

Goodman, A., Johnson, P. and Webb, S. (1997) *Inequality in the UK*, Oxford, Oxford University Press.

Hatt, S. (2000) 'The different worlds of work' in Dawson, G. and Hatt, S. (eds) *Market, State and Feminism: The Economics of Feminist Policy*, Aldershot, Edward Elgar.

Hayek, F.A. (1984) *1980s Unemployment and the Unions: The Distortion of Relative Prices by Monopoly in the Labour Market* (2nd edn), London, The Institute of Economic Affairs.

Himmelweit, S. and Simonetti, R. (2000) 'Nature for sale' in Hinchliffe, S. and Woodward, K. (eds) *The Natural and the Social: Uncertainty, Risk, Change*, London, Routledge/The Open University.

HM Government (1994) *Competitiveness: Helping Business to Win*, Cmnd 2563, London, HMSO.

Jessop, B. (1990) *State Theory: Putting Capitalist States in their Place*, Cambridge, Polity Press.

Jones, P. (1989) 'The neutral state' in Goodin, R. and Reeve, A. (eds) *Liberal Neutrality*, London, Routledge.

Keynes, J.M. (1930/1981) Evidence to the Macmillan Committee, reprinted in *The Collected Writings of John Maynard Keynes*, Vol.xx, *Activities 1929–31: Rethinking Employment and Unemployment Policies*, London, Macmillan.

Keynes, J.M. (1936/1964) *The General Theory of Employment, Interest and Money*, London, Macmillan.

Mackintosh, M. and Mooney, G. (2000) 'Identity, inequality and social class' in Woodward, K. (ed.) *Questioning Identity: Gender, Class, Nation*, London, Routledge/The Open University.

Modood, T., Berthoud, R. *et al.* (1998) *Ethnic Minorities in Britain: Diversity and Disadvantage*, London, Policy Studies Institute.

Moggridge, D.E. (1976) *Keynes*, London, Macmillan.

Pigou, A.C. (1949) *'The Economist', John Maynard Keynes (1883–1946)*, Cambridge, Cambridge University Press.

Woodward, K. (2000) 'Questions of identity' in Woodward, K. (ed.) *Questioning Identity: Gender, Class, Nation*, London, Routledge/The Open University.

Wright, R. and Ermisch, J. (1991) 'Gender discrimination in the British labour market: a reassessment', *Economic Journal*, no.101, May, pp.508–21.

FURTHER READING

Cairncross, A. (1992) *The British Economy Since 1945: Economic Policy and Performance, 1945–1990*, Oxford, Blackwell. An accessible and authoritative account of the 'golden age' of capitalism and the neo-liberal revolution, by a government economic adviser during the 1960s.

Folbre, N. (1994) *Who Pays for the Kids? Gender and the Structures of Constraint*, London, Routledge. An in-depth examination of the relation between caring for children (and other dependents) and wage employment.

Gallie, D., White, M., Cheng, Y. and Tomlinson, M. (1998) *Restructuring the Employment Relationship*, Oxford, Clarendon Press. A comprehensive and up-to-date analysis of flexible work in the UK.

Hatt, S. (2000) 'The different worlds of work' in Dawson, G. and Hatt, S. (eds) *Market, State and Feminism: The Economics of Feminist Policy*, Aldershot, Edward Elgar. An informative and accessible survey of women's employment in the UK.

Hayek, F.A. (1944) *The Road to Serfdom*, London, Routledge and Kegan Paul. The classic text of twentieth-century liberalism.

Keynes, J.M. (1936) *The General Theory of Employment, Interest and Money*, London, Macmillan. The most influential economics book of the century, and the only one that can be read and enjoyed as literature.

Moggridge, D.E. (1976) *Keynes*, London, Macmillan. This remains a clear and readable introduction to Keynes's life and thought.

Welfare: from security to responsibility?

Ross Fergusson and Gordon Hughes

Contents

chapter 4

1 INTRODUCTION

Few UK citizens live without relying on the welfare state at some point. For most of us, our health, our education and our financial security in times of unemployment, infirmity and old age depends on state-provided welfare arrangements. But what the state can and should do to maintain and improve the security of its citizens has been fundamentally challenged. Alongside changes in families and in work, these challenges have been a key source of uncertainty and insecurity at the end of the twentieth century. The costs and the principle of the state's provision of social security have been questioned. In particular, the power of the state, and its potential for ordering people's lives, both as tax-payers and as recipients of welfare, has come under scrutiny. Major changes have already occurred and more are promised. The second transformation of welfare provision in the last 50 years is under way. The first transformation was the period of radical reform after the Second World War, which brought together the institutions of the modern welfare state. Such transformations have profound effects on people's lives, their incomes, their security and even their longevity, as some of the changes in the values of old age pensions, shown in Box 4.1, illustrate.

BOX 4.1   **Changes in the value of state pensions**

The 1930s

In the 1930s, a third of pensioners lived below the poverty line (Rowntree, cited in Beveridge, 1942, para 235). A pension of 10/- (50p) per week for a single person was inadequate to meet the following outgoings:

TABLE 4.1   Requirements for retired persons at 1938 prices

REQUIREMENTS FOR RETIRED PERSONS AT 1938 PRICES

	Man and Wife	Man	Woman
Food	11/6	6/–	5/6
Clothing	2/8	1/4	1/4
Fuel, Light and Sundries	5/–	3/–	3/–
Margin	2/–	1/6	1/6
Rent	8/6	6/–	6/–
	29/8	17/10	17/4

Source: Beveridge, 1942

In addition, benefits regulations meant that families entitled to unemployment and other benefits could not afford to maintain elderly relatives at home, since the value of their pension was deducted from benefit entitlements. As a result, many pensioners lived permanently in lodgings in wretched conditions on subsistence diets. George Orwell's (1937) account of pensioners' lodgings in *The Road to Wigan Pier* describes landlords taking out private life assurance on sick pensioners as a means of making money from their deaths.

The 'golden age' of state welfare

By 1948, the value of state pensions had leapt by 160 per cent to 26/- (£1.30) for a single person. At the peak of its value, in 1978, the £31.70 pension for a married couple was the equivalent of 39 per cent of average male earnings (DSS, 1997).

FIGURE 4.1   State welfare for all
Source: *Picture Post*, 1947

The twenty-first century?

By 1998, the comparative value of state pensions for a couple had fallen below 27 per cent of average male earnings (DSS, 1999). The charity 'Help the Aged' calculated the weekly outgoings of a pensioner as £79.97, against a basic state retirement pension of £64.70. Government statistics indicated that 21,000 old people would be likely to die in that year as a result of cold-related illnesses. At the same time, there has been a rise in the number of well-off pensioners which is also indicative of the more diverse lives in the late twentieth century.

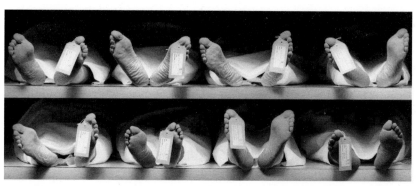

THOUSANDS OF ELDERLY PEOPLE WILL STOP FEELING THE COLD THIS WINTER

Don't let the winter kill. Call 0800 75 00 75

Help the Aged

FIGURE 4.2
Source: Help the Aged

These transformations are the focus of this chapter: how they occurred, the ideas that lay behind them, the changes they made in the ordering of everyday lives, and how power was exercised through them. In Section 2 we look at the so-called 'golden age' of security and the welfare state based on the political ideology of social democracy, concentrating on the social values that underpinned it, the institutions and structures through which it worked, and the ways in which it exerted power and ordered lives. Section 3 looks at three major critiques of the social democratic welfare state, based on the political ideologies of Marxism, feminism and liberalism. Liberalism, in particular, was a major influence on the second transformation. How it influenced change in the welfare state is the subject of Section 4. And finally, in Section 5, we consider how these transformations may be reworked to shape the welfare state at the beginning of the twenty-first century.

2   THE 'GOLDEN AGE' AND THE SOCIAL DEMOCRATIC WELFARE STATE

Most people who were born around the start of the twentieth century lived with real insecurities. If they were working class they probably lived through periods of unemployment and serious poverty. If they were ill they were unlikely to have access to treatments beyond home remedies. If they were parents they may have seen a child die before it reached adulthood. If, on the other hand, they ran a factory, owned land or helped govern the country, they would have lived amid the privileges of wealth and status but in the background was the spectre of class conflict, civil unrest or even revolution. The first half of the twentieth century was characterized by insecurity, the risk of conflict, and incipient disorder. The short period between the two devastating world wars brought a series of threats to different sections of the population. Economic depression, mass unemployment, extensive poverty and reductions in means-tested 'national assistance' benefits resulted in strikes, lock-outs, demonstrations and civil unrest during which the coercive forces of the state were required to maintain order. Conflict between an impoverished working class and an identifiable ruling class was at its peak. With it came unprecedented class consciousness, working class militancy and, for the ruling class, a fear of communism and class revolution.

FIGURE 4.3   The coercive powers of the state were required to maintain order. Hunger marchers clash with mounted police, 1932

Source: Hannington, 1936

2.1   From unrest to war, and a new social order

As little as a decade later, in the 1940s, this would have been barely recognizable as a description of the exercise of power and the maintenance of order in UK society. The threats of unrest, particularly in the form of mass civil action, had largely disappeared. The shadow of insecurity cast by mass unemployment, sickness and old age had greatly receded for most of the population. New forms of ordered living were emerging built around the kinds of job security which resulted not only from Keynesian policies of demand management (see Chapter 3) but also from greatly improved welfare rights for many people. How was such a rapid transformation possible?

A year after World War Two began, *The Times* published the following editorial:

> If we speak of democracy, we do not mean a democracy which maintains the right to vote but forgets the right to work and the right to live. If we speak of freedom, we do not mean a rugged individualism which excludes social organisation and economic planning. If we speak of equality, we do not mean a political equality nullified by social and economic privilege. If we speak of economic recon-struction, we think less of maximum production (though this too will be required) than of equitable distribution. ... The European house cannot be put in order unless we put our own house in order first. The new order cannot be based on the preservance of privilege, whether privilege be that of a country, of a class, or of an individual.
>
> Source: *The Times*, 1 July 1940

Even a year earlier an editorial like this in the national newspaper of record would have seemed outlandish, even subversive. Now it was presented as calm reason. It symbolized a shift in the ideas, the beliefs and even the social values that held sway. In other words it was a shift in the *power* of one political ideology over another. Changes like these can occur when the values and distribution of powers which prevail are no longer able to contain social divisions and conflicting interests. By 1940, the ideological positions that were used to justify the insecurity, inequality and poverty of the past by appeal to the market or the 'natural order' of things were becoming increasingly difficult to sustain. Before the war, the poorest and least secure sections of the population were relatively isolated from those with wealth and power. But during the Second World War, on the battlefield, sheltering from air-raids in tube stations, and hosting inner city evacuees, some of these barriers were confronted. The extent of inequality and its resulting injustice, at a time of supposed national unity, could no longer be defended. The legitimate power of the state to send the mass of its male working class population to possible death and probable injury was considerably weakened by pre-war class conflicts. The force of law to conscript soldiers to the front was supplemented by the power of persuasion that all would share more equally in the benefits of a free nation liberated from the threat of fascism. The dependence of the nation on working-class men and women, many of whom were deeply disenchanted with the prevailing social order, built up a debt which *The Times* was beginning to acknowledge. Power had begun to 'change sides'.

As a small step towards recognition of that debt, the Conservative-dominated war-time coalition government asked a leading civil servant, William Beveridge, to review the ramshackle system of social insurance so that it would provide greater security for those who became unemployed, ill, injured, homeless or too old to work. The Beveridge Committee in 1942 produced a plan which went much further than Churchill, the Prime Minister, had intended (Timmins, 1996). The Conservative Party's initial resistance turned to support as the weight of mass expectation of a new social order after the war became clear. Continued resistance to the plan would certainly have cost the Conservatives the 1945 election, and the Labour Party's landslide victory is partly attributed to its association with the plan, and its commitment to a transformation ambitious enough to make Britain the 'New Jerusalem'.

FIGURE 4.4   The first Labour landslide: the promise of security, Transport House, 1945

As the front page of the *Daily Mirror* (over) shows, the publication of the Beveridge Report during the war was met with euphoria. The **welfare state** which took shape after the war hugely improved the security of most of the population.

Pensions, as we saw, increased by 160 per cent; local authorities became responsible for the residential care of the elderly; benefits for sickness,

Welfare state
A state that is partly dedicated to providing for the security and well-being of its citizens.

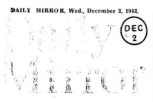

DAILY MIRROR, Wed., December 2, 1942.

DEC 2

No. 12,159 ONE PENNY
Registered at the G.P.O. as a Newspaper.

Allies separate Axis armies in Africa

AMERICAN and French forces were reported last night to have driven a wedge between the two Axis armies in North Africa—between Nehring, fighting to hold Bizerta and Tunis, and Rommel, at bay at El Aghella.

Messages from North Africa and New York stated that the Allies have reached the Tunisian coast between Gabes and Sfax.

Nehring's land communications with Tripolitania—where the Afrika Korps is preparing for the next big attack by General Montgomery's Eighth Army—have thus been cut off.

The Americans and French are believed to have pushed to the coast through desert-like country from Gafsa, in Central Tunisia.

In the Bizerta-Tunis triangle, Morocco radio reported, the British First Army has crossed the Axis minefields to come to grips with the main defences.

The French are reported to have captured Pont du Fahs, a railway town 35 miles south of Tunis.

Quit Bizerta 'Drome?

The air battle over Tunisia is being fought with an intensity believed to be unequalled since the Battle of Britain.

As the struggle grows fiercer the Germans throw in planes rushed to North Africa from Western Europe and Russia.

One Nazi pilot shot down in Tunisia was flying over Stalingrad less than a fortnight ago.

Bizerta airfield has been so devastated that it was believed at Allied H.Q. last night that most of the Luftwaffe bombers and fighters have been driven back to Sicily.

Flying Fortresses have also pounded Tunis, Sfax and Gabes.

RAF Bombers rained explosives on Bizerta during Monday night, and daylight had scarcely appeared yesterday when other Allied aircraft took up the attack.

El Agheila spearhead

Patrol activity by the advanced spearhead of the Eighth Army on the El Agheila front yesterday means that the battle for this vital position may be looming.

Indications of our growing strength in this advanced area are our constant air attacks on enemy communications on the road between El Aghella and Tripoli, and on the two enemy bases of Misurata and Tripoli.

German radio last night admitted that the spearhead of the Eighth Army had pushed closer to Axis lines, and yesterday's Italian communique also reported activity between advanced units.

F D R AND DE GAULLE

President Roosevelt said yesterday that he would be glad to see General de Gaulle, but had not invited him to visit United States.

Darlan is 'Chief of State'

ADMIRAL DARLAN has created an Imperial Council at Algiers and has assumed the powers of Chief of State in French Africa, Morocco radio announced last night.

The radio said Darlan has assumed the power of Chief of State "as representative of Marshal Petain, who is at present a prisoner."

"French Africa has resumed an official status which will enable it, pending the liberation of Metropolitan France, to defend the general interests of the Empire, to resume effectively the fight at the side of her Allies, and to represent France in the world."

The Imperial Council has already held its two first sittings.

Admiral Darlan presided over the sittings, which were attended by General Nogues, Governor - General Boisson, Governor - General Chatel, General Giraud, and General Bergeret.

STATEMENT ON DARLAN IN SECRET

The position of Admiral Darlan and the military developments in North Africa are to be subjects of a statement in secret session in the Commons.

When Mr. Eden announced this yesterday he added that an opportunity would be provided for a debate if desired.

Beveridge tells how to
BANISH WANT

Cradle to grave plan | All pay— all benefit

SIR WILLIAM BEVERIDGE'S Report, aimed at abolishing Want in Britain, is published today.

He calls his Plan for Social Security a revolution under which "every citizen willing to serve according to his powers has at all times an income sufficient to meet his responsibilities."

Here are his chief proposals:

All social insurance—unemployment, health, pensions—lumped into one weekly contribution for all citizens without income limit—from duke to dustman.

These payments, in the case of employees, would be:

Men 4s. 3d.	Employer 3s. 3d.
Women 3s. 6d.	Employer 2s. 6d.

Cradle to the grave benefits for all, including:

- **Free medical, dental, eyesight and hospital treatment;**
- **Children's allowances of 8s. a week each, after the first child.**
- **Increases in unemployment benefit (40s. for a couple) and abolition of the means test; industrial pension in place of workmen's compensation.**
- **A charter for housewives, including marriage grant up to £10; maternity grant of £4 (and 36s. for 13 weeks for a paid worker); widow's benefit; separation and divorce provision; free domestic help in time of sickness.**
- **Old age pensions rising to 40s. for a married couple on retirement.**
- **Funeral grants up to £20.**

To work the scheme a new Ministry of Social Security would open Security Offices within reach of every Citizen.

The 1d.-a-week-collected-at-the-door insurance schemes of the big companies would be taken over by the State.

Sir William says the Plan depends on a prosperous Britain, but claims that it can begin by July 1, 1944, if planning begins at once.

[See pages 4, 5 and 7]

NO COMMONS DEBATE BEFORE NEW YEAR

FIRST Commons comments on Sir William Beveridge's plan will be made at the next sitting, as the I.L.P. amendment to the Address is so broadly phrased that it will embrace any reconstruction proposals of this kind.

Mr. James Griffiths, from the Labour Front Bench, may indicate some of his party's reactions, but, as Sir William Jowitt indicated yesterday, the report will not be fully debated until the New Year.

The people of Occupied Europe are being told by radio of the report and its implications.

From dawn, in twenty-two languages, they will be shown how Britain, even in the midst of war, has grappled with social problems, just as in the past she took a lead on questions of social security.

Sir William will explain his report on the radio at 9.25 tonight.

Govt. give hint of post-war planning

CHOOSING the eve of the publication of Sir William Beveridge's long-awaited report on Social Security, the Government yesterday gave the country its first indication of its own plans for post-war Britain.

These include the continuance of rationing for some time and control of industry (some industries being taken over as public corporations); the development of agriculture, forestry and public utilities like electricity.

The Government also announced the immediate setting up of a new Ministry of Town and Country Planning, and the rejection of the Scott and Uthwatt Committees' proposals for placing main responsibility for planning in the hands of a permanent commission.

Victory First

Sir William Jowitt, Paymaster-General, answering a debate on reconstruction in the House of Commons, said:

"We must not allow ourselves to be distracted by talk of reconstruction from the stern task of securing victory.

"Talk of reconstruction is a mockery while the world is to remain hereafter under the constant fear of aggression."

Sir William referred to the Beveridge Report and said:

"The subject of Social Security is one to which all thinking men and women can subscribe. We must survey his work as part of our reconstruction work as a whole.

"I hope that early in the New Year members will be in a position to discuss the main questions raised in the Report."

Sir William said it seemed obvious that the immediate

Continued on Back Page

The Missing Link

FIGURE 4.5 From cradle to grave ...

Source: The *Daily Mirror's* front page, 2 December 1942

industrial injury and disablement rose greatly; unemployment benefits increased. The establishment of the National Health Service guaranteed free access to general practitioners and hospitals for the first time. All were entitled to free secondary schooling and subsidized school meals. The principle of **universalism** was established in some areas.

These developments produced a welfare state in which services were integrated into a relatively comprehensive programme of social improvement designed to insure all against misfortune, replacing the pre-war accumulated and limited patchwork of state assistance for some. Alongside the nationalization programmes of the major public utilities, high levels of employment and Keynesian demand management, as described in Chapter 3, the social landscape began to change. The idea of **social citizenship** came to prominence, through the recognition of social rights, equalities and expectations alongside civil and political rights. The social values of liberty and equality became more than just rhetoric, as people began to acquire the means to exercise their rights effectively. As Figure 4.6 shows, welfare spending rose and poverty declined. A 'golden age' of security appeared to be dawning in the years following the Second World War.

Universalism
The principle of entitlement to welfare for all, as a right, irrespective of means.

Social citizenship
The right and the power to be included in all key aspects of a society as an equal participant.

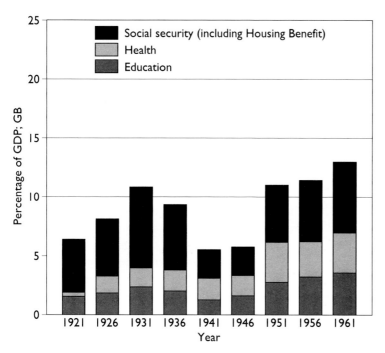

FIGURE 4.6 Welfare spending as a percentage of Gross Domestic Product (GDP), 1921 to 1961
Source: Timmins, 1996, p.522

Most of these changes in welfare and security were directly associated with the political ideology known as social democracy (introduced in Chapter 3) and the institutions that resulted are often referred to collectively as the social democratic welfare state.

2.2   Social values, institutions and power in the social democratic welfare state

2.2.1   Social values

Collectivization of risk
Pooling the costs of misfortune between all members of a society, through a programme of social insurance.

Social democracy showed a zealous commitment to welfare rights. The state was to be the guardian of the welfare of the community. At the heart of social democratic ideology is a commitment to social citizenship, the **collectivization of risk**, and the provision of security through the state. Equal social rights are intended to allow healthy, secure, educated citizens to find a place in the community and to enjoy equal *opportunities* with all other citizens to develop and prosper. These rights are matched by rights of political citizenship, principally rights of representation and political participation associated with liberal democracy. And it is by democratic means that the state is to be empowered to secure equal rights and social justice.

2.2.2   Institutions of welfare

The institutions of the state have a number of key features which proponents of social democracy see as neutral and impartial, and consistent with promoting social justice and equal opportunity. Institutions like the NHS, a school or a social services department exist independently of governments and politicians. They remain as governments come and go. They can weigh up competing interests fairly and rationally, and promote social justice and equal opportunity.

State institutions have three ways of maintaining this 'impartial' approach: specialist knowledge, professional expertise and bureaucratic procedures.

Specialist knowledge comes both from the natural and the social sciences. Medical science shows which treatments are capable of curing disease; social sciences show how to equalize opportunities or enhance social justice. Research can show why boys and girls perform differently in school (see **Gove and Watt, 2000**); it can offer explanations of extremes of wealth and poverty (**Mackintosh and Mooney, 2000**); or, as Keynes attempted, economic theory can try to make the workings of the economy as reliable as the flow of electricity (see Chapter 3). Of course, there would be considerable dispute regarding the claims of natural scientists (see **Hinchliffe, 2000**) and social scientists about the reliability of this knowledge. But their use of evidence and theory gives them a claim to being impartial and obtaining 'objective' knowledge.

Professionals
Highly qualified practitioners who have considerable autonomy within institutions and who regulate entry into their own occupational group.

The expertise of **professionals** is also built on specialist knowledge. They are represented as unbiased, neutral agents exercising judgement on the basis of knowledge and experience, which gives them the authority to decide what form of welfare is appropriate and justified. Teachers, for example, select children for special education as gifted or slow learners, social workers assess risks in violent households, and so on.

Balancing these judgements about individuals are **bureaucratic** procedures and rules that ensure consistency of practice and a fair allocation of resources. For example, the local housing officer administers a points system to ensure that the most needy are housed first; civil servants check that benefits payments accord with policies, and so on.

To the institutions of the welfare state, citizens are **clients**. They depend upon the specialist knowledge of professionals and their expertise in deciding what forms of support, treatment or protection would be beneficial. In doing so, clients accept the authority of professionals and bureaucratic administrators. That is to say, they *subject* themselves to the powers of professionals and administrators who are allowed to order clients' lives. So power is embedded in the structures of welfare institutions, and enacted through the agency of professionals and administrators.

Bureaucracy
A predictable, rational system of administration for organizing and controlling the work of institutions.

Client
Someone in a dependent position who seeks professional protection or assistance.

2.2.3 Welfare, power and ordering lives

A C T I V I T Y 4.1

Think about your last contact with a welfare state institution, such as an appointment with a doctor, attending a job centre, meeting a child's teacher or collecting a pension. Are there any ways in which it could be said to have ordered your life or the lives of others? Did it involve the exercise of power?

C O M M E N T _____

It is easy to see how welfare institutions impose order on people's daily lives. The timing of the school day or appointments to meet doctors or social workers or the geography of surgeries and hospitals can become fixed points around which some people's lives are constructed. Less obviously, welfare agencies shape other aspects of people's actions. For example, your GP might be urging you to lose weight. To remain eligible for the job-seekers' allowance, the job centre might be requiring you to attend interviews for jobs that do not attract you. Your child's teacher might be prevailing on you to stop him bullying others. The Post Office counter clerk may be making you feel uncomfortable that a well-dressed, car owner draws a state pension.

_____

All these situations have the potential to make welfare clients *act in ways which they would not otherwise have chosen* – eating less, considering low-paid work, scrutinizing how a child plays with others, choosing a different Post Office counter. In very simple ways, the client has been changed by his or her dependency on the institution. In other words, the client has fulfilled the condition that Chapter 1 defined as demonstrating the exercise of power.

Thinking back to the 'modalities of power' outlined in Chapter 1, what kinds of power are being exercised in the above examples?

The examples include authority, coercion, domination, persuasion and manipulation. The GP may have been *persuading* you, but she does so by being '*an authority*' whose advice you take seriously. This time she is *in authority* – using her powers to deploy the resources of the NHS. At the job centre there is no choice but to go through a series of regulated activities. This is close to *domination*: you must submit to these procedures if you want your allowance. The parent whose child is accused of bullying is experiencing *normative pressure*, in which someone in authority appeals to shared understandings about acceptable behaviours. Behind that pressure is the unspoken *coercive* threat of exclusion from school. The affluent pensioner is feeling stigmatized as an undeserving recipient of welfare, a form of *domination* in which one party appeals to prevailing values to undermine another.

All institutions of welfare exert power like this. Their rules, their objectives and the actions of their administrators and professionals shape lives. Their structures define the extent of their power. The actions of their agents seem to match the conscious and deliberate exercise of *hierarchical* power, by someone 'over' some else, suggested by Weber's conception (as outlined in Chapter 1). On behalf of the state, people are recipients of welfare in ways that shape how they live, sometimes profoundly. Professionals, for example, place pupils in special units for the disturbed; social workers commit people to detention in psychiatric units; children are taken into care.

In fact, many of the workings of the institutions of state welfare fit Weber's theory. Power is often *top-down*: the welfare state comprises a structure of institutions whose capacity to exert power is enshrined in law. The scope of state power is extended by its *delegation* through thousands of welfare institutions. This is based on *authority*: the legitimacy of institutions' actions is partly derived from popular support, partly from their legal status, and partly from the authority of expert agents. Much of it is also based on claims to *rationality*: power is exercised by reference to 'scientific' evidence about who deserves benefits, how resources for schools and hospitals are best deployed, and so on. *Impartial* rules determine what actions should be taken, irrespective of who is the subject of a welfare intervention or who is administering it. *Bureaucracy* ensures impartiality, it protects the recipient of welfare from the personal intrusions of the officer giving it, and it protects the officer from the actions and feelings of the recipient. This may be achieved by locating contact either literally in offices, a kind of neutral territory, or by assigning the provider with 'an office' in the sense of a role or title (housing officer, employment adviser) which gives them another *guise*. Officers are active *agents* who know how far they are permitted to draw upon the authority of their office to discipline, treat or support welfare subjects.

Weber viewed the state as an independent source of power in its own right, with a range of means for exerting power built into its institutions, from the coercive powers of the police and the military to arrest offenders or disperse strikers, to the legal powers of professionals to sedate the mentally disordered. In this sense, the state is seen as holding the balance of power between competing interests – social classes for example – on the basis of its own rationality.

Foucault's theory, as described in Chapter 1, sees welfare institutions as ordering lives in quite different ways. Power works within the human imagination, shaping what it is possible to think and talk about, and what it is not. It is not that thoughts or ideas are forbidden, but that they do not arise, because other ideas have centre stage, or because they do not fit recognized ways of thinking. Familiar chains of language and connections between observations gradually come to define what is known and accepted. So the knowledge and thinking that shapes an institution is the product of collective views built up in layers over years of history. It is gradually moulded to fit the life and culture of the institution and is normalized as 'natural' or inevitable. In the examples we looked at, agents acting on behalf of the institution are not so much 'following orders' as reproducing for others, as truth and common sense, a set of values and norms they have internalized from their own exposure to the culture and structures of the institution. Your GP can 'trade' with you about your body weight because you allow yourself to be positioned as someone she talks to like this: you *collude* in giving her power. The conduct of the employment adviser or the Post Office counter clerk is not so much using explicit authority as adopting 'ways of being' which they 'picked up' from others when they were learning their technical skills. In this sense, their power is contingent and open to challenge.

These are two very different understandings of power at work, but in both it is clear that lives are being ordered by institutional power in the processes of organizing and dispensing welfare.

ACTIVITY 4.2

On the grid overleaf, note:

- the key social values of social democracy
- the key concepts used by social democracy in explaining the workings of the institutions and structures of the welfare state
- the key concepts used by social democracy to explain how the welfare state exerts power and orders lives.

Social values	Institutions and structures	Power and ordering lives

SUMMARY

- The period before the Second World War was characterized by insecurity for much of the population, and by conflict between social classes.

- The Second World War marked a shift in political ideologies, and a transformation of prevailing values in favour of reducing inequality and creating greater security.

- The welfare state established an extensive programme of welfare provision and increased security during this period.

- Rising welfare expenditure and a reduction in the proportion of the population living in poverty led to the view that this was a 'golden age' of welfare.

- The post-Second World War welfare state is closely associated with the political ideology of social democracy.

- The key social values of social democracy are those of equality of opportunity, social justice and social citizenship.

- Social democracy looked to the state and its institutions to realize these values, regarding it as a neutral structure capable of allocating resources equitably, under democratic control.

- The institutions of state welfare exert power and order lives through the exercise of professional authority, bureaucratic regulation, and the application of supposedly impartial rules in a 'rational' system, according to Weber; or through the normalization of particular ways of understanding the world, according to Foucault.

3 POLITICAL IDEOLOGIES: VALUES, INSTITUTIONS AND POWER

. . . .

According to social democracy, the welfare state orders lives fairly and impartially, in the interests of the whole of society, and secures a more equitable distribution of resources and power. The 'golden age' of welfare and Keynesian full employment greatly increased opportunities for all citizens. But how power works in and through the institutions of the welfare state, and for what purposes, is seen very differently by other political ideologies. To make sense of the transformation of the welfare state at the end of this 'golden age', we need to understand why different political ideologies — Marxism, feminism, liberalism — took a much more critical view of the power exercised by welfare institutions.

Each of these ideologies has several variants, and is also associated with a long history of social science thought. The following outlines sketch only the broad fundamental ideas of each ideology, in terms of:

- the social values on which they are based

- how they see the institutions and structures of welfare

- their interpretation of how power works to order people's lives.

3.1 Marxism

Social democracy places great emphasis on the power of agency to effect change, from the power of intellectuals like Keynes, to the actions of state officials. But as we saw at the end of Chapter 3, some political ideologies stress the importance of structures. According to Marxism, the inequalities of the 1920s and 1930s were the direct result of the class structure. The **capitalist mode of production** exploits the labour power of the working class by placing its members in subordinate positions. Because they do not own land or property and can live *only* by selling their own labour, they do so at low prices that are fixed by powerful employers. The real value of their labour, reflected in the market worth of what they produce, is not returned to those who created it, but is expropriated as profit by owners. What has given value to the goods is not the material itself, but the human labour added to it to convert it from raw material freely available in the environment, into usable items. But what workers are paid is the lowest wage the labour market will allow, *not* the value that they have added to cheap raw materials. By taking the difference as profit, the owners exploit the workers. This causes the conflict between workers and employers we saw in Chapter 3, which is

Capitalist mode of production
A system based on private ownership and control of production for profit, using hired waged labour.

inherent in the *structure* of production in capitalist societies. It makes society fundamentally unequal and prone to unrest and incipient disorder, such as occurred in the 1920s and 1930s. To Marxists, the welfare state was the self-interested concession of a beleaguered bourgeoisie, designed to stem the tide of mass discontent (see also **Mackintosh and Mooney, 2000**).

Social values

This analysis informs the core social values of Marxism. Throughout history, a wealthy controlling elite has exerted its overwhelming power to exploit the repressed mass of the population: brute force was used to exploit slaves; control over land allowed the exploitation of serfs; the ownership of the means of production gives the bourgeoisie power over the proletariat. In each case, it is the labour power of the repressed class that is exploited. So, for Marxism, the core values of its politics are the equalization of power, and freedom from exploitation. When this is achieved, the working class will realize its creative potential to work co-operatively, experience fulfilment in work, and enjoy the fruits of its endeavours. This can only occur in a society in which social equality prevails.

Institutions and structures

For Marxists, far from making capitalism less unjust, the welfare state protects it by disarming criticism and diluting discontent. By offering some security of material conditions and rights to health and education for some, it dispels the claim that social class at birth determines future life-chances. By establishing the possibility of social mobility, it gives legitimacy to a social order that is divided by class. It also creates an educated, healthy workforce who are ready to generate more profit and reproduce the next generation of workers ready for exploitation. So long as living standards are rising, demands for radical change to the social order will accumulate little support.

Marxists also argue that the welfare state was a highly pragmatic response to difficult circumstances, as the Conservative Party quickly recognized during the Second World War when it abandoned its opposition to the Beveridge plan.

Power and ordering lives

According to Marxism, the power of the state to tax, redistribute wealth, and repress or dispel opposition is used to maintain a class divided society. The state is a tool of the dominant class and it always acts in this class's interests. In the case of state welfare, by paying a relatively small price through taxation, capitalists can continue to accumulate profit without fear of unrest. The kinds of rationality and expertise on which the institutions of state welfare draw are not 'scientific' or 'objective' but are constructed by an emerging professional class whose thinking is shaped by the social values and aims of the dominant class. The resentment of the exploited proletariat is 'managed' away by the millions of small, everyday actions by which doctors, teachers, social workers, social security officers and others order people's lives. The casualties of capitalism 'benefit' enough from state welfare to prevent them becoming the discontented working class which so alarmed governments before the Second World War.

ACTIVITY 4.3

On the grid below, note:

- the key social values of Marxism
- the key concepts Marxism would use in criticizing the institutions and structures of the welfare state
- the key concepts of Marxism's critique of how the welfare state exerts power and orders lives.

Social values	Institutions and structures	Power and ordering lives

3.2 Feminism

Social values

Feminism's dispute with the social democratic welfare state settlement shares much in common with that of Marxism. Like Marxism, feminism also sees this state as consolidating unequal relations of power, namely those of men over women. Far from equalizing opportunities for women, much of the workings of the welfare state have served to lock women into domestic 'housewife' and 'carer' roles, limit their educational and career horizons and keep them out of the labour market. As we saw in Chapter 2, one social value of feminism is that of equality between men and women, which in turn leads to an emphasis on the idea of social justice with regards to financial independence, citizenship rights, paid employment and social welfare. Some feminisms give particular emphasis to the positive recognition of difference and diversity with regard to people's living arrangements, their sexuality and their capacity to develop themselves as they choose without male oppression. This recognition of difference and diversity brings feminism into dispute with social democracy's repression of difference and diversity in the living arrangements institutionalized and idealized in the old welfare state (i.e. the 'natural' nuclear family described in Chapter 2). Feminism also celebrates the unpaid and low-status caring work of women, viewing it as a key feature of a true welfare society. Different feminisms (as outlined in Chapter 2) place different emphases on these values but all versions identify the institutions of state welfare as an historical source of women's oppression.

Institutions and structures

As Chapter 2 also showed, the family is regarded as a – for some *the* – critical institution within which the unequal relations between men and women are set and maintained. Women's financial dependency, their historical exclusion from paid work, their concentration in low paid and less secure employment, their responsibility for domestic work and caring for others all stem from the family. And it is mainly within the confines of the family that their sexual exploitation and oppression occurs, and where hidden violence, or the threat of it, maintains male power over women. Many of the criticisms of the social democratic welfare state raised by feminism are, therefore, about its support for the patriarchal family. The family is the key unit through which welfare is organized and dispensed and gender inequalities institutionalized. Welfare professionals identify and treat individuals principally as members of families. Women are seen first as part of a unit dominated by men, rather than as individuals. Their needs are judged in relation to their families' needs.

The subordination of women's position through the family was consolidated by the very reforms for which the social democratic welfare state became celebrated. The idea of 'the family wage' for example, was fundamental to the Beveridge plan. It overtly viewed women's economic security as a by-product of men's. The male breadwinner was enshrined in the benefits legislation of the so-called 'golden age'. Women were seen as wives and mothers, rebuilding a strong nation through childbirth and maternal care. The right to paid employment was implicitly challenged by claims that young children needed the attention of their mothers, not trained carers in nurseries. Through schools, hospitals, GP surgeries, midwives, benefits offices, social work teams and an array of other welfare state institutions, these messages were consolidated.

In an era when the family has changed profoundly as an institution for ordering people's lives, this picture appears outdated. Most women are in paid employment. There have been massive increases in the numbers of women who live alone, and of lone mothers. Women's financial dependency on men for purposes of family allowances, taxation and pension rights has been reduced greatly. The pressures, especially on lone mothers, to take employment, and the growth of child care provision to allow it, have overturned the idea of the home-bound mother caring for children. But as feminist arguments showed (see Chapter 2), many of these changes impose a dual burden of work and domestic labour on women. The fundamental assumptions of difference between men and women, and their translation into unequal powers and rights, are seen to have simply taken more modern forms.

Power and ordering lives

How are these powers that keep women in weak, dependent positions maintained? In particular, how do tens of thousands of unconnected employees of welfare institutions appear to 'act in concert' and order families in ways that are not required by policies or fixed in law? As Chapter 2 showed, Foucault's theory of power helps to make sense of how male dominance becomes normalized in family life. Part of this normalization takes the form of the ordinary knowledge of everyday life. But these normalizations are strengthened by the kinds of expertise to which welfare professionals and administrators lay claim. The knowledge and thinking that shapes the actions of an institution, such as a social work department or a doctor's surgery, is the product of collective understandings of people who hold positions of power, built up in layers over years, and gradually moulded to fit the life and culture of the institution. Ideas that become embedded in this way appear to have an existence in their own right. So, for example, the belief that only women can be effective primary carers of infants persists, despite the evidence that men can (and do) care for children successfully.

Complex networks of norms and assumptions accumulate around welfare institutions. In millions of everyday interactions, values and 'truths' are reproduced. Welfare institutions thus have the power to order family lives, and the place of women within them, through taken-for-granted practices and everyday transactions.

ACTIVITY 4.4

On the grid overleaf, note:

- the key social values of feminism
- the key concepts feminism would use in criticizing the institutions and structures of the welfare state
- the key concepts of the feminist critique of how the welfare state exerts power and orders lives.

Social values	Institutions and structures	Power and ordering lives

3.3 Liberalism

Liberalism's dispute with the welfare state is quite different. As we saw in Chapter 3, in recent decades it has been the driving force for flexible working. It has also profoundly shaped the efforts to transform the welfare state.

Social values

Liberalism's core values are to guarantee the autonomy of individuals, to facilitate social diversity and to ensure state neutrality. Individual rights and freedoms are paramount, especially the right to accumulate private property and enjoy the rewards of personal enterprise and risk taking. This is a driving force of progress; it not only benefits the individual but also the larger society. In return for individual gain, there must be individual responsibility. People must provide for themselves and their families, not look to others for help.

The contrast between these values and those of social democracy is stark. Liberalism focuses on the individual; social democracy on the social. Liberalism tends to favour personal gain which increases social inequalities; social democracy favours equality of opportunity. Liberalism wants minimum interference in people's activities; social democracy looks to an active state to intervene. Liberalism wants to make individuals and families responsible for their own welfare; social democracy sees responsibility as lying ultimately with the state on behalf of society as a whole. From these starting points, liberalism is hostile to much of what the welfare state stands for.

Institutions and structures

As we saw in Chapter 3, for liberalism, the key institution for holding society together and securing these values is the free market. The welfare state is seen as undermining the operation of the free market. By collectivizing risk though social systems of insurance the welfare state imposes high levels of taxation which penalize most the very people who make provision for themselves. Not only is this an infringement of people's fundamental right to own property and accumulate wealth, it is also a disincentive to enterprise. Taxation is the act of an over-active state using legal powers to redistribute wealth from the enterprising, the independent and the successful to the dependent and the idle. Furthermore, according to liberals, giant bureaucracies like the National Health Service waste taxpayers' money in planning and administration. In a free market, the complex work of co-ordinating activity and matching the supply of a service to demand for it is undertaken with immense efficiency by the 'invisible hand' of the price mechanism (see **Himmelweit and Simonetti, 2000**). Competition ensures that private firms find inventive ways of minimizing the costs of administration and maximizing the quality of service they provide. Rather than trying to provide welfare services, the state should confine itself to ensuring that the market operates fairly and efficiently, by restricting monopoly providers for example. At most, the state should provide a minimal safety net for those who are so disadvantaged they are incapable of surviving in a market system.

Power and ordering lives

To liberals, what holds society together and orders lives is people's interdependency, and the contracts they make with each other in a market system. These are not just contracts about buying and selling goods or labour power, but about how to behave and co-exist. In the economy, the market creates the 'spontaneous order' described by Hayek in Chapter 3, ordering lives through the way employment is shared out. Within the structures of free markets, people have individual agency to decide how to behave – whether to work hard and take multiple flexible jobs, balance their work life against leisure pursuits, and so on.

Who has power and who does not, for liberalism, depends largely upon success in the market. Those who make most profit gain most power. While this means a very unequal distribution of power, that is inevitable in societies marked by great diversity and differences in cultures and abilities. It is, nevertheless, a just system because power is allocated on an impartial competitive basis, not by patronage.

Once again, the contrast with the forms of ordering envisaged by social democracy, particularly through the welfare state could not be greater. There, lives are ordered by institutions, professionals, and bureaucratic rules and structures based on expert knowledge and rational systems. To liberalism, all these methods quash creativity, reward weakness and depend upon constant state intervention to secure order. In the attempt to make the distribution of power less unequal, power becomes concentrated in the hands of unelected bureaucrats and self-interested professionals.

ACTIVITY 4.5

On the grid below, note:

- the key social values of liberalism
- the key concepts liberalism would use in criticizing the institutions and structures of the welfare state
- the key concepts of liberalism's critique of how the welfare state exerts power and orders lives.

Social values	Institutions and structures	Power and ordering lives

COMMENT

When you have completed this grid for liberalism, check your version of the tables you have completed for Activities 4.2 to 4.5. against Table 4.3 at the end of the chapter.

3.4 Social exclusion and the welfare state

Criticism of the social democratic welfare state does not come solely from other political ideologies. The claim that the welfare state made universal provision is challenged by a wide range of intellectuals and pressure groups representing excluded groups. They see the development of welfare within the UK in an international and historical context, partly in terms of how the UK formed itself as a nation and a colonial power. They view the consensus that established the welfare state as part of a larger settlement, in which the nation came to see itself in a particular configuration of social relations, constructed around a conception of 'Britishness' as gendered, white, able-bodied and heterosexual. Williams' (1989) analysis of the **social settlement** argues that at the end of the Second World War, the UK had to come to terms not just with class inequalities but also with a colonial history, as well as questions of who and what constituted the nation, and norms regarding women that were increasingly being challenged. The connection between welfare entitlement and nationality had been established since the 1900s. The unfolding sense of a modern nation and social citizenship after 1945 had to take account of the war-time debt to ex-colonies and the mass immigration of black British citizens who were brought to the UK to fill gaps in the labour market – not least in the poorest paid jobs in the welfare service. Migrants were excluded from welfare partly by residency rules and partly through cultural forms of racism. Similarly, bureaucratic rules and dominant norms positioned many women and people with disabilities outside the scope of effective welfare provision. Equal social citizenship was not extended to those who did not 'fit' the social settlement.

Social settlement
The conception of social relations on which the welfare state was built, based on ideological assumptions about and connections between nation, family and work.

4 RESTRUCTURING THE WELFARE STATE

This chapter began with the transformation in the organization of welfare following the Second World War. Forty years later a second transformation changed lives significantly. By the 1980s the values and practices of social democracy were challenged by social scientists and politicians who were strongly influenced by liberalism. A number of factors combined to bring this about, many of which, as Chapter 3 showed, also brought Keynsian economic policies into question. They included protracted conflicts between workers and managers, high inflation, rising unemployment, cheap imports from a globalizing market, claims of poor productivity and lack of competitiveness within industry and, especially, criticism of welfare spending. The costs of welfare, its adverse impact on work and enterprise, and the emergence of a 'culture of dependency' were all challenged. This section looks at this second transformation and the attempts made to restructure the welfare state along liberal lines.

4.1   Restructured institutions, re-ordered lives?

ACTIVITY 4.6

Think about one area of state welfare that you know, and make notes on major changes you noticed during the 1980s and 1990s. As you think about the changes, think, too, about what seemed to be the purpose of them.

COMMENT

You may have noticed changes in everyday things like names on signs, with terms such as 'grant-maintained school' and 'NHS Trust Hospital' starting to appear. Perhaps you were directly affected by changes, for example by a local hospital closure or buying your council house; if you were unemployed you may have found yourself re-named as a job seeker; maybe your GP moved to a group surgery located further away from you.

Whatever the points you noted, you probably identified the main purpose of these changes as being to save money.

These examples represent significant changes to the way in which welfare state institutions were organized. Some state schools ceased to be the responsibility of local councils and were funded direct from central government. NHS hospital trusts became separate entities responsible for their own budgets. Councils were forced to sell council houses. Entitlement to long-term unemployment benefit was reduced and claimants were required to demonstrate that they were actively seeking work. Fund-holding GPs grouped together to cover each other's work and make the budgets they controlled go further. Old people in need of long-term nursing care ceased to be entitled to free local council provision if they had substantial capital assets. School meals were contracted to commercial caterers.

In one important sense the main purpose of these changes was indeed to save money or, more accurately, to shift the burden of the cost of services to relieve the tax-payer. There was a drive for greater efficiency and for more competition to cut costs. This accounted for many of the changes in health care; for example NHS trusts were forced to specialize and compete with one another to drive down the costs of operations.

But these were not the only purposes of the changes. They were also intended to give a greater say to those who used welfare services, and greater responsibility and control to those who managed them. Control over schools shifted from local councils to school governors and included strong parent representation. Head teachers gained greater power to shape their schools, but also had greater responsibility for any shortcomings.

Changes were also intended to promote greater responsibility of individuals and families for their own lives, and to reduce reliance on the state. Individuals were responsible for finding jobs, not the state for providing them. Well-off older people had no need of tax-payers' subsidies for residential nursing care. There was a drive away from standard state-issue provision, to promote choice and to meet the needs of an increasingly diverse population. Council house buyers looked forward to changing the appearance of their home to suit their own taste rather than the local council's.

The connection between these aims and the ideas of liberalism is unmistakable. The commitment to the freedom of the tax-paying consumer is clear. Allowing free choice, fostering diversity, and expecting responsibility run through these changes. So does minimizing the role of a controlling state. The drive to reduce taxation is intended to return control to the individual. Empowering managers is intended to promote efficiency through competition, and so on.

The liberal inspired Conservative governments of Margaret Thatcher and John Major between 1979 and 1997, then, aimed to transform the old social democratic welfare state into a less costly, smaller, more efficient and more flexible form, capable of meeting basic needs and diverse preferences, while shifting more of the burden of cost and responsibility to individuals and families whenever possible. At the same time, the users of welfare institutions were intended to be made more powerful, along with the people who manage them, and the power of the experts and professionals who provided the services was to be contained to allow this.

Did these plans of restructuring succeed? The rest of this section tries to answer this question by asking two further questions:

- Did the costs of the welfare state decline when the 'golden age' ended?

- Has the power of the users and of the managers of welfare services increased, and has the power of experts declined?

4.2 Efficiency, economy and effectiveness: rolling back the social democratic welfare state?

ACTIVITY 4.7

Have a look at the following statements and at Table 4.2 and ask yourself what these statistics appear to show.

Between 1981/2 and 1995/6, total expenditure on social welfare increased by 62 per cent in real terms

Unemployment expenditure increased by 21 per cent from £7.5 billion in 1981/2 to £9.0 billion in 1995/6

Social security expenditure due to family breakdown rose from £2.1 billion to £9.5 billion (345 per cent increase) from 1981/2 to 1995/6

Expenditure on the elderly increased by 36 per cent over this period

(*Social Trends*, 1997)

TABLE 4.2 General government expenditure in the UK: by function, £ billion at 1995 prices

	1981	1986	1991	1994	1995
Social security	61	76	83	100	102
Health	26	29	35	40	41
Education	28	29	33	37	38
Defence	25	29	26	25	23
Public order and safety	9	10	15	15	15
General public services	9	10	13	13	14
Housing and community amenities	14	12	10	11	10
Transport and communication	8	6	8	–	9
Recreational and cultural affairs	3	4	4	4	4

Source: adapted from *Social Trends*, 1997, p.118, Table 6.21

COMMENT

The answer is fairly simple isn't it? The outcomes show some dramatic increases (look at social security for example), zero saving across social welfare expenditure and generally increased total expenditure.

So increased spending occurred in this period despite the sustained ideological and political attack on the size and features of, and expenditure on, state welfare throughout the decades of the 1980s and 1990s. Much of the welfare state therefore appears to be quite resilient to any root and branch change. This leads us on to a more complex question that we need to try to answer. How might we explain this seemingly contradictory result of sustained ideological and political assaults on the welfare state and its institutions *not* producing economic savings by the 1990s? To help us with this conundrum, think back to the claims made in Chapters 2 and 3 about the changing patterns of lifestyles, households/families, and work and (un)employment. What did we learn about the likely economic consequences of these demographic and work-related changes in the last decades of the twentieth century? High levels of unemployment, changing and increasingly diverse family forms and lifestyles all carry welfare implications. The triumph of liberalism in 'rolling back' the state by reducing the extent and costs of the interventions is therefore limited.

The explanation of this seeming conundrum of very high welfare costs persisting after attempts to cut them lies in the increased expenditure in areas of public spending associated with the decline of full (male) employment, as well as changes in the demographic profile of the UK in this period. Unemployment benefits necessarily rose sharply given the unacceptability of leaving the involuntarily workless without any means of economic subsistence. The growing proportion of the population living well beyond retirement age also meant that increased expenditure on older people's health and care needs was politically necessary. Such countervailing factors help explain the seeming 'failure' of the attempt to roll back the state's expenditure on social welfare.

The answer to our first question in this section, 'did the costs of the welfare state decline when the 'golden age' ended?' appears to be no. This does not mean that nothing has changed in the ways in which our lives are ordered through social welfare. Let us now examine our second question: 'has the power of the users and of the managers of welfare services increased, and has the power of experts declined?'

4.3 Changing cultures and power relations in the restructured welfare system

In Section 2 above, we drew a picture of the old welfare state in which the expert, whether a professional or bureaucrat, ruled supreme. Expertise was certain given both its rule-bound rationality and scientific mystique. Now we have a great deal of questioning of experts across the board, from food scares based on BSE and GM foods, to the 'failures' of doctors and social workers in the institutions of welfare. Have we increasingly lost faith in experts' proclaimed capacity for ordering our lives in 'our' best interests? At the same time, we have also seen the development of a consumer culture in welfare provision.

In this section we focus on the rise of two new figures in the emergent culture of the restructured welfare system:

Welfare manager
Co-ordinator and promoter of welfare production modelled on the company manager in the private sector.

Welfare consumer
Term used to capture the shift from passive recipient to active choice-maker in relation to welfare services.

- the **welfare manager** as the new expert replacing the bureaucrat and the professional
- the 'active' **welfare consumer** replacing the dependent/passive client of the old social democratic welfare state.

It has been widely claimed that both these figures break with the old monopoly power and expertise of professionals and bureaucrats and associated assumptions about passive and uninformed 'clients', said to characterize the old social democratic welfare state. It is certainly true that the new forms of privatization and the introduction of market mechanisms have changed the routine, institutional relations of welfare provision. The following questions help us explore these cultural and organizational changes in greater depth:

1 How does the power of welfare managers differ from that of the old social democratic welfare experts, and how are institutions of social welfare re-ordered through these changes?

2 If 'we' are now welfare consumers rather than clients or citizens of social welfare, what are the implications of this new 'identity' for the ways in which people's lives are getting re-ordered at the end of the twentieth century?

From bureaucrats and professionals to managers?

There have always been managers in both the private and the public sectors. However, there is currently much greater prominence given to the role of managers as new experts in social welfare. If you work in the public sector or know someone who does, you may have noticed the shift from the talk of 'public servants', 'professionals', 'civil servants' to that of, for instance, 'purchasers', 'providers' and 'managers'. The expertise of the manager resides chiefly in his or her capacity to make social welfare organizations more 'businesslike' and more customer-centred, with the manager striving for better resource control and thus greater efficiency. Of course, we are not suggesting there has been a total break with the past ways of organizing welfare and its power relations based on bureaucratic, professional,

FIGURE 4.7 Let's play accountants and fund holders, Unison advertisement, 1994

'scientific' principles and practices (see Section 2.3 above). Rather, it is likely that professional experts (such as doctors) remain important players – an instance of continuity with the past – but there are also new dominant means of ordering things and people, as epitomized by the emergence of such figures as the clinical director, business manager, unit manager, chief executive and such like in hospitals. These practices and priorities are often termed **managerialism**.

Managerialism
The belief in enhancing the power of managers to re-order public sector institutions more efficiently, effectively and economically.

ACTIVITY 4.8

Look back to Section 2.2.2 and the discussion of the bureaucratic and professional forms of expertise in the old social democratic welfare state. Try to summarize how such experts were held accountable for their decisions.

COMMENT

Accountability in bureaucracies is upwards through the hierarchical organization and is based on a clear set of rules that all office holders are required to follow. In professions, accountability is to fellow professional colleagues and to a professional body. In the institutions of the social democratic welfare state there is a mixture of these forms of accountability.

Managerial accountability by contrast is based on the accountability of staff to the organization for which they work *and* to the customers or consumers of the service. Much of the attraction of managerialism lies in its apparent dynamism and innovative potential for challenging the 'old' bureaucracy and its professions associated with seemingly obsolete and inefficient patterns of service provision. This process seems to involve a dispersal of power through devolution and decentralization: policy makers clarify goals and outcomes and managers are left to get on with the job through their agency. Managers are thus 'set free' to go about their business, so, for example, in social care, 'care managers' put together and control packages of care for service users. You came across this claim that an identifiable shift has taken place in the way power 'descends' through the social welfare hierarchy in Chapter 1. In that chapter, John Allen suggested that recent developments in social welfare have seen the decline of the direct, bureaucratic structures of supervision and control etc., and their replacement by new front line agencies and their managers who are held in check by monitoring and contractual arrangements exercised 'at a distance' by the centre. The role of the state as the primary provider of social welfare has thereby declined. Its activities and power have increasingly been transformed and dispersed into that of being the *enabler* for a range of diverse organizations, families, communities and individuals in the production and distribution of social welfare. Such developments both enhance the power of the centre but also open up greater possibilities for the distortion and ambiguity of decision-making practices by the freed-up front line managers. For example, managers are encouraged to 'be enterprising and innovative' and to 'take risks', the downside of which may be a departure from the principles of probity and fairness and the established rules and regulations set up to protect both staff and service users.

From citizens and clients to consumers

So what has happened to the 'client-citizen' of the old social democratic welfare state services? We have already suggested that the 1980s and 1990s saw the emergence of a new social identity around that of being a customer or consumer of welfare services. Let's examine what the term 'consumerism' implies for an understanding of the changing power relations around social welfare. Talk about consumerism concentrates on the idea of the autonomous individual protecting his or her own personal interests (or those of family members) in the context of guarding against risks to welfare and security. In the language of consumerism, the market is recognized and celebrated as the most efficient means by which to realize individual choice and diversity, and to respond to particular needs. Being a welfare consumer thus opens up considerable promise for 'agency' not least by heralding a shift from the 'we know best' of the professional and 'do as you're told according to our rules' of the paternalistic bureaucrat to a new language and practice of choice, responsiveness and quality. The connections between this shift in language and the liberal critique of the social democratic welfare state (see Section 3.3) will be evident. Here we should note the growing appeal to the 'knowledgeable' welfare consumer in place of the old figure of the 'client-citizen'. This consuming individual is also viewed as acting knowledgeably and powerfully in and *through* the new flexible welfare markets.

Both the shift towards managerialism and the emergence of the new welfare consumer represent important changes in the way power is distributed and lives are ordered through the institutions of state welfare. Managerialism, as we have already noted, brings a shift towards dispersed forms of power which, as Chapter 1 suggested, often end up being unstable, unreliable, diffuse, and a very long way from the tightly controlled chain of command of Weber's theory. As John Allen hints, this style of highly dispersed local decision making in diverse institutions, at times run by private and voluntary providers, comes closer to Foucault's 'hit and miss domination'. In this version, individual managers as semi-autonomous agents are powerful and work in networks which go beyond their state 'bosses'. Of course, past practices and financial and administrative structures put firm limits on their freedom but they do have the scope to negotiate and organize their services to order lives differently.

In the jargon of liberalism, like managers, new welfare consumers are said to be 'empowered', by being released from the old structures of standard 'one-size-fits-all' state welfare. If we follow Foucault's theory, the ways of thinking about being the dependent client have been challenged. The old routines and procedures which made citizens passive recipients of services decided by professionals and bureaucrats were gradually challenged by erosions of authority, questioning of expertise and new ideas. Accepted ways of thinking about needing medical treatment or an income in old age began to shift from acting as a dependent to being responsible for one's own security. Clients-turned-customers were being shaped by a new set of anonymous forces into 'governing the self' in a different way.

We can see that Weber's and Foucault's theories offer ways of understanding this transformation in the way people's lives are ordered through welfare. But it is important to recognize that they continue to vie with each other to explain both the 'golden age' of welfare and the restructured forms from the 1970s to the 1990s. Superficially, Weber's theory looks as though it fits the highly regulated old structures of the 'golden age' and Foucault's the more diverse and changing forms of 1990s welfare. But it would be a mistake to get drawn along this line of thinking. Foucault's theory would dispute Weber's explanation of institutional power in the 'golden age'. It would argue that the institutions of welfare always had only partial success in imposing order through rules, procedures, expertise and so on. Similarly, a Weberian theorist would want to argue that despite some superficial changes and increased freedoms for managers and consumers, the underlying distribution of power in the organization of welfare, and the social outcomes it produces remain largely unaltered – in other words, hierarchical power relations and structured inequalities remain to the fore. One important feature of this latter claim would be to ask for whom the claimed freedoms of the new welfare consumer are real and for whom they are no more than rhetoric. For the latter group, inequality would be an important indicator of old power structures remaining intact. The political ideologies of Marxism and feminism would support this claim.

New welfare, diversity and inequality

New forms of welfare provision have partly been inspired by the need to recognize the growing diversity of the population, and the increasingly diverse ways in which lives are ordered. Accepting that the welfare needs of black women are as legitimate as those of white men, for example, has partly informed the 'freeing up' that the changes brought. But the question of how the distribution of resources works alongside diversity is also important. Many commentators now distinguish between 'old' poverty, caused by old age or illness – in other words, circumstances that can and will affect al! of us – and 'new' poverty, which is the particular experience of groups vulnerable to marginalization. As **Mackintosh and Mooney (2000)** show, income in the UK remains very unevenly distributed. But it is clear that inequalities have been increasing.

BOX 4.2   **Groups most vulnerable to poverty**

These include:

- Women, who are much more likely to be poor than men. This has to do with the fact that women still retain primary responsibility for home and family.

- Unemployed and low-paid people.

- Lone parents, the majority of whom (93% in Scotland) according to the 1991 Census are women.

- Rural households with low income and poor access to public services.

- Young people – 16 and 17 year olds – who have no job, no Youth Training place and no benefits.

- Disabled people or families with a disabled child. The additional costs of caring and limited access to the labour markets make this group of people particularly susceptible to poverty.

- Ethnic minorities who have much higher rates of unemployment and are disproportionately represented in low paid jobs.

- Families with children: children increase the costs of essentials, and this at a time when one parent (usually the mother) stops work to care for them.

- Pensioners who are dependent on state benefits or small occupational benefits.

Overall, between 1979 and 1993 the number of people in the UK living on incomes below half the average income rose from 5 million (9 per cent of the population) to 13.7 million (24 per cent of the population).

Source: Scottish Poverty Information Unit, 1997

The effects of this departure from the 'golden age' of the social democratic welfare state are arguably that of dividing the life-chances of the population more dramatically along the inequalities of class, 'race', gender, able-bodiedness and age than ever since the 1930s, and reversing the greater equality trend of the period 1945 to mid 1970s. As Marxist commentators would note, for the privileged consumer there may be more choice and power, greater healthiness and well-being, but for the poor non-consumer, trapped in poverty, there are probably worsening life-chances and less choice. Thus greater opportunities and choices for the relatively privileged welfare consumer appear to accompany greater risks and uncertainties for the poorer sections of UK society whom Bauman (1997) has ironically termed 'flawed consumers'.

FIGURE 4.8   Homeless children in a hostel: poverty in the UK in the 1990s

It is likely that feminist critics would also point to the differential resources and opportunities between men and women for consumer choice. The idea of the welfare consumer does appear to register diversity in that everyone has different wants but it tends to downplay the inequalities of social differentiation. 'Playing' the role of a welfare consumer is not a 'game' carried out on a level playing field. The odds are differentially distributed according to the resources on which the person is able to draw. Indeed there is evidence of growing inequalities linked to class, 'race', age, gender, able-bodiedness and place.

Since the early 1980s there has been a renewed emphasis on the responsibilities of 'working' and so 'consuming' households and families providing for their future security against risks and uncertainties through private insurance of various sorts. This approach goes hand in hand with the changing role of the state whereby it no longer assumes responsibility for collective risk management in a direct fashion. Alongside this there has been a growth in what we may term **privatized prudentialism** by the individual consumer with regard to risk management (O'Malley, 1992).

To take the example of private health insurance, in 1955 in Britain there were 300,000 subscriptions to private health insurance compared to 3.6 million (covering 6.2 million people) in 1996 (Brunsden, 1998, p.155). The private provision of residential care for older people has also grown in significance during the last decades of the twentieth century. Whereas £6 million of public money was spent on this in 1978, by 1990 it was closer to £1000 million (Allsop, 1998). This may be said to involve the commercialization of need whereby the responsibility for meeting needs shifts increasingly from the state

Privatized prudentialism
Strategy whereby individuals are exhorted by government to be responsible for and to calculate privately their own risks and opportunities.

to individuals and families – as welfare consumers – in the new markets in care. Here we are witnessing new ways of being governed, in Foucault's terms 'governing our selves'. Once again, this development has resulted both in new inequalities and very different degrees of uncertainty and insecurity between differentially resourced groups as well as more specialized modes of care for those able to choose a design for life from the diverse range of services on offer.

Let's now try to make sense of these specific developments in terms of what they mean for the state's attempts to order our diverse lives. At the end of the twentieth century, individuals and families are increasingly exhorted by governments to be responsible for, and to privately calculate, their own risks and opportunities in the realm of welfare. The increasingly dominant rationale for social welfare is one based on the responsible welfare consumer. It marks a significant departure from the social democratic collective management of risk based on compulsory contributory insurance schemes organized by the state discussed in Section 2.

SUMMARY

In this section we have focused on:

- the new imperatives of cost-cutting and competition in the context of welfare restructuring

- the continuing high levels of state expenditure on welfare despite the attacks and critiques followed by explanations for these seeming contradictions

- the recognition, nonetheless, of key cultural and organizational changes associated with new expertise and authority of managers and the rise of welfare consumerism.

There are also important continuities with the past. Structural inequalities and different patterns of vulnerability based on class, gender, 'race', age, able-bodiedness and place, remain marked and, arguably, have become more acute since the 1970s.

5 ORDERING LIVES AND WELFARE IN THE TWENTY-FIRST CENTURY

FIGURE 4.9   The second Labour landslide: the promise of security in return for responsibility, Downing Street, 1997

In this chapter we have explored two major post-Second World War transformations in the way the organization of welfare has ordered lives. Section 2 examined the main features of the social democratic transformation and the new power arrangements associated with this welfare state. Section 3 then examined the main criticisms lodged at the feet of the 'old' social democratic welfare state, emphasizing the key role played by liberalism in both the ideological assault on social democracy and the subsequent political programme of welfare reform. In Section 4 we examined the main contours and consequences of this restructuring of welfare. This restructuring has not been without its critics and has been subject to contestation, not least around the dangers of leaving social welfare to the wishes of managers, consumers and markets with the consequence of the growing inequalities for the most vulnerable sections of society. So far, then, this chapter has focused largely on two sharply contrasting 'ways' of organizing and ideologically justifying different programmes of social welfare. The 'first way' we may term social democracy and the 'second way' that of liberalism. Since the mid 1990s, however, there has been a growing interest in attempts to formulate a 'third way' which, supposedly, keeps the best features from the first two but provides a middle way between the two extremes of social democracy and liberalism. Does such thinking represent the beginning of a third transformation of welfare as we enter the twenty-first century? We cannot hope to answer this question here but it is important for us to begin to explore some of the emergent features of the changing direction taken by welfare policy in the late 1990s. For our purposes the social policy project of the Labour government elected in 1997 provides a useful case study through which to examine both change and continuity in the provision and rationale of social welfare. We can also explore the extent to which its proposals depart from the values associated with the political ideologies that you have encountered in this chapter.

By the end of the twentieth century there appears to be widespread dissatisfaction with both (a) the emerging and growing welfare inequalities associated with the restructured system of social welfare since the 1970s and (b) the seemingly unstoppable costs of welfare. At the same time there is now a dominant consensus across different political parties and welfare agencies that 'good management' is the secret to how best to run welfare organizations and to solve the economic problems of the welfare state.

Section 2 above has already alerted us to the close relationship between political ideology and social democracy and the development of policy making in the welfare state. Furthermore, in Sections 3 and 4 we have also noted the pronounced influence of liberal ideas on the restructuring of the welfare state since the 1970s. Since the election of the Labour Party into government in 1997, there has been another example of the close interplay between academics and intellectuals and politicians. This has been widely known as 'The Third Way' or the 'new' Centre-Left. In 'think-tanks' such as Demos, academic centres such as the London School of Economics (Giddens, 1998) and directly in pamphlets and speeches by leading politicians (Blair, 1998), this approach has been celebrated both as a break with 'old' political traditions and ideologies and as a modern approach 'fit for purpose' for the new times of greater uncertainty, diversity, opportunity and flexibility. Let's examine the main claims of this latest agenda of transformation in social welfare.

ACTIVITY 4.9

Read the following press release from the Department for Education and Employment in May 1999 following a lecture by the then Education and Employment Secretary, David Blunkett.

In this lecture Blunkett argues that the government's welfare reform programme was about finding a better way than either the 'rampant individualism' (of liberalism) or the 'debilitating welfarism' (of old social democracy). The real challenge was to enable people to become self-reliant, in keeping with the spirit of Beveridge's original objectives, rather than forever dependent on benefit.

There are two issues on which we'd like you to focus while you read the extract:

1 Think about the assumptions it contains with regard to how the state should shape and attempt to order lives 'from above' and how far welfare should be about us 'governing ourselves' as responsible agents.

2 You should also think about which influences from the political ideologies explored in this chapter are evident in this celebration of the responsible, working citizen. (You might want to refer to Table 4.3 at the end of the chapter.)

> This is the first time since the birth of the modern welfare state that any Government has been prepared to take such a clear-headed approach to reform – recognising that every person who can do so should have the chance of working and looking after themselves and their families, rather than being dependent on benefits for long periods of time. In our drive to modernise the welfare state we face those on one side of the political spectrum advocating rampant individualism, and on the other, well-meaning people calling for dependence based on debilitating welfarism. The traditions of self-help in this country were built on inter-dependence and not dependence on the state. Those who tried through mutual support to help themselves sought Government backing to enable them to do so, not to remove their involvement in the decisions that affected their lives.
>
> How do we ensure that these values underpin the changes we need to make to the Welfare State for the 21st Century when so many see such change as a threat rather than a reassertion of more traditional values and solutions? ... We want to return to the driving force which created the welfare state – self-help through mutual help and not state welfare dominated by benefit dependency. Resources should be targeted at providing the means for individuals to flourish through learning, through work, through security at times of change and dignity in retirement or severe disability.
>
> *[The speech then outlines the successes of the New Deal for young people in providing people 'the opportunity' and means to escape a life of benefit through work or training as a route to work. Blunkett then notes that there are some abusers of the scheme who do not wish to work and proposes removal of benefits for six months for those sanctioned three times.]*
>
> Not working is not an option ... This is 'tough love', which doesn't punish but offers help in return for endeavour. This is the challenge for a Government genuinely taking on the task of transforming welfare into work and dependency into hope for the future.
>
> (Department for Education and Employment, 1999)

C O M M E N T

1   This is clearly a vision of the state that involves an attempted ordering of lives from above. And yet you may also have noted that it is also calling on us to govern ourselves as responsible *and working* individuals.

2   The message of the lecture appears to contain a complicated mix of liberal and social democratic values.

Let's unpack these points in a little more depth. The lecture appears to be calling for us all to become more responsible for our future well-being, if you like 'governing ourselves' in the Foucauldian sense. The call for greater responsibility for the individual, particularly through a revamped work ethic, appears very close in its values to liberalism. But it is also clear from the thrust of this lecture that it is the government or the state making the running on this agenda for re-ordering our lives. Indeed there appears to be the attempted imposition of new norms and practices of work and welfare from above, with the overt threat of state coercion if people do not co-operate in the new 'tough love' agenda. Here we can see the relevance of both Foucauldian and Weberian theorizing of power. In terms of the role of the state it is claimed that it is neither that of social democracy's 'debilitating welfarism' (too much state intervention) nor the anti-state 'rampant individualism' of liberalism (too little state intervention).

We have already noted the close affinity of some of the proposals contained in the lecture to liberalism's stress on individuals' responsibility for self-help and self-reliance through the work ethic ('Not working is not an option'). Blunkett clearly wishes to criticize the tendency for the old welfare state to institutionalize benefit dependency on the state rather than autonomy for the individual. However, we can also see close links back to social democratic values, not least in the celebration of the government's leadership role in creating the conditions for the New Deal which 'offers help in return for endeavour'. Finally there is a striking similarity to the claims of the post-Second World War social democratic consensus that it had found the middle way between socialism and capitalism.

6 CONCLUSION

In this chapter we have traced two major transformations in the provision of state welfare in the era that followed the Second World War. The first transformation consolidated the position of the state as the key to universal welfare provision, in pursuit of the social democratic aims of creating security for all. This so-called 'golden age' of welfare improved provision for many and achieved genuine security for some. Thirty years on, the beginnings of the second transformation were visible, and by the 1990s a substantive shift had occurred from the premise that the state could, and should, ensure the social security of its citizens, to an assumption that individuals and their families have responsibilities for providing for their own security. The transformation was particularly visible in the major changes in the way recipients and providers of welfare were understood. Welfare consumers and service managers began to replace the old conceptions of the client-citizen and the welfare professional and bureaucrat.

Both transformations, as we saw, represented crucial changes in the ways the institutions of state welfare ordered people's lives. The client-citizen of the immediate post-war years was understood as the compliant, dependent subject, entitled to welfare as of right but expected to accept the prescriptions of welfare professionals and officials. In extreme cases of dependence or containment, lives were literally 'ordered' by the decisions of state welfare employees. The life of the welfare consumer who emerged in the 1980s and 1990s was ordered in significantly different ways, exercising degrees of choice, and even some power as a consumer, but as a responsible agent who lives with the consequences of his or her choice and who is enjoined not to see the state as the ultimate source of his or her well-being. In these ways lives have been re-ordered, both in the sense of changing how the institutions of state welfare cause us to act, and in deciding who gains and who loses.

These transformations, we have argued, were profoundly shaped by two political ideologies. Much of the thinking that drove the welfare reforms of the 1940s and 1950s came from social democracy, which sees an extensive welfare state as the key means to achieving many of the social objectives regarding citizens' rights, security and equality of opportunity. The transformations of the 1980s and 1990s drew heavily on the social values and the political and economic objectives of liberalism, particularly regarding individual freedoms, personal responsibility and the containment of the role of the state. But other political ideologies also play a key role in making sense of how the institutions of state welfare order lives, both in terms of the critiques they have made of social welfare, and the influence of their ideas in shaping policies. Feminism, as we saw, tends to view state welfare as having historically consolidated the dominance of the patriarchal family, but it also informed policy changes that limited this dominance. Meanwhile, Marxism stresses the key contribution of state welfare to perpetuating the exploitative relations of production which are inherent in all capitalist societies.

Throughout this chapter, we have seen that the exercise of power is clearly about ordering people's daily lives. But exercising power also produced the changes in social relations that our stories of transformation in state welfare have highlighted. In particular, we have employed two very different theories of power and how it works in and through the institutions of state welfare, associated with the ideas of Weber and Foucault. It is clear that there are Weberian understandings of the old modes of ordering around the client-citizen and the professional expert and office-holder. But the Weberian approach to power is also helpful in exploring the re-ordering of welfare subjects as consumers and managers in the 1980s and 1990s. And, as we have also seen, both moments of transformation in the lives of welfare subjects are also open to interpretation from the Foucauldian approach to power. Each theoretical approach therefore takes us in important but different directions in the study of the two major transformations in state welfare.

Where, then, is welfare heading in the twenty-first century?

Section 5 above has helped us explore some of the key features of policy initiatives on welfare at the end of the twentieth century. We cannot yet say whether this new approach represents a real break with the second transformation or whether it involves a revisiting of 'old' social democracy. At the time of writing there remains much uncertainty about the likely success of government proposals for welfare reform. As we have noted throughout this chapter, structures of power are always contested, provisional in character and unstable in their outcomes. Things change and planned outcomes don't always happen as intended. Rather than indulge in speculative predictions of the future, we have aimed more modestly to introduce you to some features of the latest *attempt* at re-ordering welfare at the end of the twentieth century. Nonetheless, the discussion should help you to understand and critically interrogate new ideological claims and political programmes, and their social values that will continue to resonate in the first years of the twenty-first century and beyond.

REFERENCES

Allsop, J. (1998) 'Health care' in Alcock, P., Erskine A. and May, M. (eds) *The Student Companion to Social Policy*, Oxford, Blackwell.

Bauman, Z. (1997) *Postmodernity and its Discontents*, Cambridge, Polity Press.

Beveridge, W. (1942) *A Report on Social Insurance and Allied Services*, London, HMSO.

Blair, T. (1998) *The Third Way: New Politics for the New Century,* London, Fabian Society.

Brunsden, E. (1998) 'Private welfare' in Alcock, P. Erskine, A. and May, M. (eds) *The Student Companion to Social Policy,* Oxford, Blackwell.

Department for Education and Employment Press Release No. 227/99, 19 May 1999

Department of Social Security (1999) *Abstract of Statistics for Social Security Benefits and Contributions and Indices of Prices and Earnings 1998,* Newcastle on Tyne, DSS.

Giddens, A. (1998) *The Third Way,* Cambridge, Polity Press.

Gove, J. and Watt, S. (2000) 'Identity and gender' in Woodward, K. (ed.) *Questioning Identity: Gender, Class, Nation,* London, Routledge/The Open University.

Hannington, W. (1936) *Unemployed Struggles 1919–1936,* London, Lawrence and Wishart.

Himmelweit, S. and Simonetti, R. (2000) 'Nature for sale' in Hinchliffe, S. and Woodward, K. (eds) *The Natural and the Social: Uncertainty, Risk, Change,* London, Routledge/The Open University.

Hinchliffe, S. (2000) 'Living with risk: the unnatural geography of environmental crises' in Hinchliffe, S. and Woodward, K. (eds) *The Natural and the Social: Uncertainty, Risk, Change,* London, Routledge/The Open University.

Hughes, G. and Lewis, G. (eds) (1998) *Unsettling Welfare: The Reconstruction of Social Policy,* London, Routledge.

Mackintosh, M. and Mooney, G. (2000) 'Identity, inequality and social class' in Woodward, K. (ed.) *Questioning Identity: Gender, Class, Nation,* London, Routledge/The Open University.

O'Malley, P. (1992) 'Risk, power and crime prevention', *Economy and Society,* vol.21, no.3, pp.251–68.

Orwell, G. (1937) *The Road to Wigan Pier,* London, Victor Gollancz.

Scottish Poverty Information Unit (1997) *Defining Poverty,* Briefing Sheet No.1 Glasgow.

Social Trends, London, HMSO (annual)

Timmins, N. (1996) *The Five Giants: A Biography of the Welfare State,* London, Harper Collins.

Williams, F. (1989) *Critical Social Policy,* Cambridge, Polity Press.

FURTHER READING

George Orwell's novel *The Road to Wigan Pier* (first published in 1937) gives a graphic description of life in Britain before the establishment of the social democratic welfare state.

An accessible and lively history of the welfare state in post-war UK has been written by Timmins, entitled *The Five Giants: A Biography of the Welfare State*, 1996, Harper Collins.

On the changing contexts of social welfare past and present which also pays detailed attention to the specific histories of different sites of social welfare, an accessible edited text is Hughes and Lewis' *Unsettling Welfare: The Reconstruction of Social Welfare*, 1998, Routledge/Open University.

There is a growing body of work on the welfare project of 'New' Labour, amongst others you may find Bill Jordan's *The New Politics of Welfare,* 1998, Sage, both lively and challenging, especially given its comparative perspective.

TABLE 4.3   Summary of political ideologies

	Social values	Institutions and structures	Power and ordering lives
Social democracy	Equality of opportunity Social justice Social citizenship Welfare rights Democratic control	Neutral state Universalism Collectivization of risk Democratic procedures Arbitrate competing interests	Institutions of state Bureaucratic power Professional expertise Structures of power Rational authority Visible power
Marxism	Social equality Freedom from exploitation Co-operation	Welfare provision dispels dissent Institutions of welfare legitimize inequalities and reproduce workforce	Structures of power Capitalist state acts in interests of dominant class Professionals and experts as agents of capital
Feminism	Gender equality Social justice Citizenship rights Recognition of difference and diversity Valuing work of carers	Welfare state legitimizes female subordination and dependency Family as key site of women's oppression	Structures of power Welfare state acts in the interests of patriarchy Experts normalize male dominance/female oppression
Liberalism	Freedom of individual Tolerance of diversity Property rights Reward of enterprise Personal responsibility	Monopoly state provision High taxation and expenditure Lack of consumer choice Inefficient public services	Weakens individual agency Over-regulates individual activity Limits self-determination Compromises the power of markets and competition Imposes particular kinds of order

Afterword

Gordon Hughes and Ross Fergusson

We began this book by asking what holds societies together – in other words, how do societies and people's lives become ordered? This overarching question was explored by examining the workings of some key social institutions in UK society following the Second World War. Throughout the four chapters of the book, we have discussed how our lives are shaped, and how people relate to each other, in terms of the exercise of power in social institutions. But we have also seen that the processes of ordering are not fixed. Our 'story' has been one of profound transformations in *and across* the institutions of the family, work and welfare in the UK over the last half of the twentieth century.

Within this overarching narrative of transformation, Chapters 2, 3 and 4 have also engaged directly with the claim that, in the last decades of the twentieth century, we have witnessed the end of a 'golden age', involving a shift from 'tradition' (Chapter 2), 'certainty' (Chapter 3) and 'security' (Chapter 4) to the new times of 'diversity', 'flexibility' and 'responsibility'. Of course, we have not suggested that these three aspects of the 'golden age' were ever entirely 'golden'. In some important ways these chapters have shown us that there never was a totally fixed order in the past, never mind a romantic 'golden age' when all was certain, secure and harmonious for everyone. The supposed 'golden age' in each institutional site was marked by its own distinct relations of power, for example class, gender and 'race' divisions were embedded in the structures of the old institutions of the family, work and state welfare. And relations of power were also marked in the norms and assumed ways of the cultures of these 'golden age' institutions. As social scientists, we need to be sceptical of any nostalgia for supposed 'golden ages' which idealizes the past.

The new institutional arrangements of family, work and welfare at the beginning of the twenty-first century reflect changed distributions of power in both their structures and processes. There is also greater uncertainty and diversity in the ways our lives are now being ordered. Throughout the book, from our opening example in Chapter 1 of genetically modified foods, to the changes in family, work and welfare, we have confronted an institutional world that looks increasingly uncertain and in flux. We have found that those features of our lives that once looked certain and taken-for-granted turn out to be 'up for grabs' in many ways, not least as a result of changing arrangements of power. What does it mean now to be a father or a mother, a worker, an unemployed claimant, or a welfare recipient? These old, taken-for-granted 'facts of life' are increasingly subjects for debate and public controversy. And yet, looking at these same institutions, it may be argued that they still have the characteristics of a social institution that we defined in the Introduction to this book: they are predictable, governed and they order lives. Being in a family, however diverse its form; being in employment, however flexible; or being a welfare recipient, however responsible a consumer, all have constraining consequences on the degree of agency that we may exercise. Children in families today continue to be socialized into the collective norms and mores of a shared culture and learn the ground rules of what it is to be a member of a human society. In turn, employment (or the lack of it) still largely governs if not always where we live then certainly what we can afford to spend our money on and how we are seen in the eyes of other people. And forms of social welfare, such as health care, remain sources of support and succour in periods of vulnerability and risk.

We can see from the above discussion that ordering continues to be tied to the work of the key social institutions in contemporary society. Although this enables us to capture the continuities that exist with the past workings of social institutions in the 'golden age', it is important to note that *how* such ordering occurs is markedly different in many ways from that which characterized the period in our society some fifty years ago. Throughout the book we have highlighted the theme of *uncertainty and diversity* as a means of helping us explore the transformations both within and *across* family, work and welfare in UK society following the Second World War. Depending on one's ideological viewpoint, we have seen that these developments may be either a source of concern and anxiety (the overturning of the stable and ordered life of the past) or a source of celebration and optimism (the greater potential for an active engagement on our part in taking up identities in the context of choice and diversity). Remember that the titles of Chapters 2 to 4 plot the possible historical move from one (old) state of affairs to that of a (new) context. In the institutional site of the family, we plotted the shift from the so-called 'traditional' nuclear family, with its tightly constrained gender roles for men and women and its taken-for-granted domestic and parenting arrangements, to more diverse ways of living together and bringing up children. What is at issue here is not the 'disappearance' but rather the profound diversification of family lives. Meanwhile the institution of paid work has witnessed the declining significance of the certainties of the 'job for life', secure employment and the reliable 'family wage' for men and, in the last two decades of the twentieth century, the growing importance of flexible and less secure working patterns and labour markets for women and men. Finally, in the context of social welfare, the extensive welfare state of the immediate post-Second World War period and its comprehensive system of state-run security has been restructured. In part, this has been replaced by a mixed economy of welfare in which personal responsibility and private provision for well-being have come increasingly to the fore. And it is vital to note that these specific transformations in each institutional site are *inter-connected* in profound ways.

Let's now re-visit the inter-connected nature of these transformations that we touched on in Chapters 2 to 4. In our discussion in Chapter 2 of changing family life in the UK following the Second World War, much emphasis was placed on contextualizing both the 'traditional' nuclear family and the more recent diverse types of family forms in terms of the dominant patterns of paid employment, past and present. The very structure of the nuclear family, with its distinct gender roles of male breadwinner and female housewife, was predicated on the particular social democratic and patriarchal settlement of both Keynesian economic policy and the Beveridgean welfare 'deal'. In turn, Chapter 2 suggested that the relative decline of the 'traditional' nuclear family is, in part, the result of the transformations in patterns of paid employment resulting in much greater female participation in work outside the home and the changing distribution of power that this may entail within the domestic sphere. Other examples of the connections between changes in our chosen institutions readily come to mind such as the influence of the political ideology of liberalism in transforming both the worlds of work and welfare since the 1970s.

The inter-connected transformations across the major institutions of contemporary societies have been seen by some commentators as indicative of a total loss of order. For example, the sociologist Manuel Castells has argued (somewhat apocalyptically) that 'Our societies are not orderly prisons, but disorderly jungles' (Castells, 1997, p.300). Furthermore, he suggests that many of our once key institutions and organizations are now 'empty shells', being increasingly unable to relate to people's lives and values. This last verdict appears to have some relevance to the more uncertain times that we now inhabit. However, we would suggest that the inter-connected, institutional transformation stories that we have explored speak more of a re-ordering of lives rather than the profound state of disorder to which Castells alludes. The latest ordering then is not just about change but also about the new patterning of things.

And where are we going from here? In part it is a truism to say that we live in changing times since this implies that the past was somehow set in aspic and was an unchanging order. However, there is strong evidence from our discussions within this book (and elsewhere in this book series) that we are living in a society that is marked by a greater sense of uncertainty and greater opportunities for diversity than was the case, say, fifty years ago. And, of course, uncertainty and diversity co-exist. As Giddens has noted, industrial societies have entered a period where 'the world is cut loose from its moorings in the reassurance of tradition' (Giddens, 1991, p.176). As we saw at the end of Chapter 4, it would be foolhardy to try and predict the future developments of social welfare. Remember our cautionary note that structures of power are always provisional and are, not least as a result of ideological contestation, unstable in their outcomes. The same note of caution applies to the institutions of the family and work. However, once again it is evident that there are clear, inter-connected developments across the three institutional sites. There is a common call from governments for us to govern ourselves as 'responsible' agents in our family lives, work and employment and in social welfare. If we fail to do so, then the state will take steps to make sure that we do. For example, of late, many governments have emphasized the need for developing parenting skills among those who do not act responsibly in family life, for the re-socialization of people to make them workers 'fit' for the challenges of the new flexible labour markets, and for the development of more prudent and self-reliant consumers of welfare services and pension schemes. Both Weber's and Foucault's views on the workings of power seem to have a resonance here. This latest attempt at the re-ordering across institutions at the end of the twentieth century remains profoundly 'unfinished business'. Nonetheless, our discussion of political ideologies will provide you with a lasting framework for making sense of the wider sets of ideas that lie behind specific policies and institutional transformations, while recognizing that policies are often hybrids, made up of different ideological strands and compromises between competing political influences. Finally, it is also crucial to bear in mind the fact that developments in the UK in general and in the operation of its key institutions, whether family, work, welfare or beyond (for example, the media, leisure, religion, politics, etc.), cannot be grasped adequately without recognizing the wider context of transnational, possibly global trends. **Held (2000)** looks at this issue in much greater depth and discusses the ways in which social institutions are being re-shaped in a 'globalizing' world.

References

Castells, M. (1997) *The Power of Identity*, Oxford, Blackwell.

Giddens, A. (1991) *The Consequences of Modernity*, Cambridge, Polity Press.

Held, D. (ed.) (2000) *A Globalizing World? Culture, Economics, Politics*, London, Routledge/The Open University.

Acknowledgements

Grateful acknowledgement is made to the following sources for permission to reproduce material in this book.

Chapter 1

Text

'Whose choice is it anyway?', *The Guardian*, 4 June 1998, © Guardian Newspapers Ltd., 1998.

Figures

Figures 1.1 and 1.10: Don McPhee/*The Guardian*; Figure 1.2: Adrian Arbib/Still Pictures; Figure 1.3: Meikle, J. (1999) 'Four new GM test sites revealed', *The Guardian*, 17 August 1999, © Guardian Newspapers Ltd., 1999; Figure 1.5: Jorgen Schytte/Still Pictures; Figure 1.6: Bildarchiv Preussischer Kulturbesitz; Figure 1.7: Marc Garanger © Editions Gallimard, Paris; Figure 1.9: Paramount/Courtesy Kobal.

Chapter 2

Text

Phillips, M. (1997) 'Death of the Dad', *The Observer*, 2 November 1997, © The Observer 1997; 'I concertina my interface with the children', *The Observer*, 11 October 1998, © The Observer 1998; 'It's cracking us up. I'm so very, very tired', *The Observer*, 11 October 1998, © The Observer 1998; Norton, C. (1999) 'Costly childcare "keeps mothers out of work"', *The Independent*, 29 May 1999.

Figures

Figure 2.1(a): Kurt Hutton/Hulton Getty Picture Collection; Figure 2.1(b): Hulton Getty Picture Collection; Figure 2.2: © Posy Simmonds. From *The Observer*, 25 October 1998. Reprinted by permission of the Peters, Fraser and Dunlop Group Ltd.; Figures 2.3, 2.4 and 2.5: Office for National Statistics (1999) *Social Trends*, no.29, © Crown copyright is reproduced with the permission of the Controller of Her Majesty's Stationery Office; Figures 2.6 and 2.7: Sally Greenhill/Sally and Richard Greenhill Photo Library.

Tables

Tables 2.1, 2.2 and 2.3: Office for National Statistics (1999) *Social Trends*, no.29, © Crown copyright is reproduced with the permission of the Controller of Her Majesty's Stationery Office.

Chapter 3

Text

'Too busy earning a living to live', *The Observer,* 11 October 1998, © The Observer 1998; Bowley, G. (1998) 'Flexibility cuts both ways', *Financial Times Weekend,* 28/29 November 1998.

Figures

Figure 3.1: Clive Branson, *Selling the Daily Worker Outside the Projectile Engineering Works,* 1937, © Private Collection; Figure 3.2: Hulton Getty; Figure 3.4 (top, left and right): Mike Levers/The Open University; Figure 3.4 (bottom): A scene from *Working With Care,* animated video, © Leeds Animation Workshop.

Tables

Tables 3.1 and 3.2: adapted from Dex, S. and McCulloch, A. (1995) *Flexible Employment in Britain: A Statistical Analysis, Research Discussion Series No.15,* Equal Opportunities Commission/*Labour Force Surveys,* Office for National Statistics, © Crown copyright is reproduced with the permission of the Controller of Her Majesty's Stationery Office.

Chapter 4

Text

Editorial from *The Times,* 1 July 1940, © Times Newspapers Ltd. 1940.

Figures

Figures 4.1 and 4.3: Hulton Getty Picture Collection; Figure 4.2: Courtesy of Help the Aged/Target Direct; Figure 4.4: Popperfoto; Figure 4.5: *Daily Mirror,* 2 December 1942, © Mirror Group Newspapers Ltd.; Figure 4.6: Hills, J. (1993) *The Future of Welfare: A Guide to the Debate,* Joseph Rowntree Foundation; Figure 4.7: Courtesy of Unison and BMP DDB Advertising Ltd.; Figure 4.8: Craig Easton/*The Independent,* 15 October 1998; Figure 4.9: Russell Boyce/Popperfoto.

Tables

Table 4.1: Cmd 6404 (1942) *Social Insurance and Allied Services,* Report by Sir William Beveridge. © Crown copyright is reproduced with the permission of the Controller of Her Majesty's Stationery Office; Table 4.2: Office for National Statistics (1997) *Social Trends,* no.27, © Crown copyright is reproduced with the permission of the Controller of Her Majesty's Stationery Office.

Cover

Image copyright © 1996 PhotoDisc, Inc.

Index